45 Structure and Bonding

Göran Wendin

Breakdown of the One-Electron Pictures in Photoelectron Spectra

With 69 Figures and 3 Tables

Springer-Verlag
Berlin Heidelberg New York 1981

ISBN 3-540-10584-0 Springer-Verlag Berlin Heidelberg New York
ISBN 0-387-10584-0 Springer Verlag New York Heidelberg Berlin

Library of Congress Catalog Card Number 67-11280

Typesetting and printing: Schwetzinger Verlagsdruckerei. Bookbinding: J. Schäffer OHG.
2152/3140-543210

Breakdown of One-Electron Pictures in Photoelectron Spectra

Göran Wendin

Institute of Theoretical Physics, Chalmers University of Technology, S-41296 Göteborg, Sweden and Department of Physics, Brookhaven National Laboratory, Upton, L.I., NY 11973, USA

The purpose of this review is to describe some spectacular effects of many-electron interactions in certain core levels in heavy atoms, as well as to give an overview of closely related phenomena in molecules, solids and adsorbates. The central concept will be what we shall call *giant Coster-Kronig (gCK) fluctuation and decay* of a hole level, involving ineraction of a single hole with configurations with primarily two holes and one excited electron. In systems with an open "valence shell" structure, i.e. with empty levels spacially as compact as the occupied ones, the interaction process can become extremely strong and lead to a breakdown of the one-electron picture. In some cases, there is even a complete breakdown of the quasi-particle picture, in which case the spectral strength has no prominent discret features and rather shows a continuum-like distribution.

The breakdown of the one-electron picture can be associated with symmetry breaking and localization, in a wide sense, of the hole. Due to the gCK fluctuation process (configuration interaction), a hole has to be described in terms of a wave-packet of one-electron symmetry states. As a consequence, in a number of cases, an atomic hole cannot be confined to a proper subshell but will move in a polarized subshell. In molecules, a hole will often not be confined to a molecular symmetry orbital but will be localized to varying degrees. Finally, in a metal, a hole in a narrow band often cannot be described in terms of extended states, in which case the non-validity of a ground state band picture may show up as band narrowing or shifted band structure. Particularly spectacular effects of localization occur in the case of two holes in a narrow band.

Examples of atomic levels showing very strong many-electron effects are $4\underline{s}$, 4p-like holes in Cd to Gd and $5\underline{s}$, 5p-like holes in Bi to Pu, where the gCK process takes the form $ns \leftrightarrows np^5nd^9nf,\varepsilon f$ and $np^5 \leftrightarrows nd^8nf,\varepsilon f$, with n=4 and 5, resp. These spectra represent a partial or complete breakdown of the quasi-particle picture. In molecules one has found a large number of cases of partial or complete breakdown of the quasi-particle picture in the inner valence region of e.g. N_2, N_2O_4, C_2H_2, CS_2, to mention only a few examples. This type of behaviour seems to be the rule rather than the exception due to the high density of two-hole-one-electron levels in the inner valence region. Finally, in metallic Ni a hole in the 3d-valence band has been found to show pronounced effects of localization, giving rise to a narrowing of the 3d-band in comparison with the calculated ground-state band structure.

Table of Contents

1 Introduction

One of the simplest elementary excitations one might create in an atomic system is the core hole left behind in an X-ray photoionization process. In an X-ray photoelectron spectrum (XPS) the core hole is usually observed as a single sharp line and this is the experimental proof of the existence of a stable core level and the validity of the concept of a one-electron orbital. In many-body language, such one-electron like excitations are called *quasi-particle excitations*. Also, sharp core levels are needed for observing sharp level structures and well defined absorption edges in optical absorption spectra. In recent years, however, it has become clear that many atomic and molecular systems do not always behave in this simple manner but show *pronounced many-body effects* even to the point of *complete breakdown of the quasi-particle picture* and the one-electron orbital concept. The most spectacular example observed so far for free atoms is the missing $4\underline{p}$ core level in Xe and surrounding elements but in addition there are a large number of related problems in atomic, molecular and solid state physics which can be explained within the same framework. The original purpose of the present article was to discuss the physics behind the disappearance of the $4\underline{p}$ core level but in view of the general character of the phenomenon it seems equally important to formulate the general framework and stress the unifying aspects. I shall therefore also give a general discussion of the various many-electron interaction processes contributing to the photoelectron spectrum and point out the conditions for having a very strong interaction and breakdown of the quasi-particle picture.

The first experimental observations of the absence of core levels seem to have been made by Lukirskii et al.[1)] and Codling and Madden[2)] in soft X-ray absorption spectra for Xe. They found lines converging to what seemed to be the $4\underline{p}_{3/2}$ threshold but no lines were found that could be associated with the $4\underline{p}_{1/2}$ threshold. Later photoabsorption experiments on Te[3)], Xe[4)] and Cs[5)] confirmed these results for Xe and Cs and showed that in Te both $4\underline{p}$-thresholds were absent (see also Sect. 9).

Direct evidence for breakdown of the one-electron orbital picture was first found by Siegbahn et al.[6)] in the X-ray photoelectron spectrum (ESCA) for I, Xe and Cs where the expected $4\underline{p}_{1/2}$ line was replaced by some very broad structure. Later, Gelius[7)] presented a much more detailed ESCA spectrum for Xe, showing a strong line which was interpreted as $4\underline{p}_{3/2}$, an extended region with pronounced fine structure and a very prominent continuum at higher binding energies. Furthermore, relativistic Δ SCF calculations[7)] failed to predict the position of the assumed "$4\underline{p}_{3/2}$" ESCA line as well as a number of other lines. For the "$4\underline{p}_{3/2}$" and $4\underline{s}_{1/2}$ lines the discrepancy was found to be as much as ~10 eV.

Gelius and coworkers[7, 8)] and Kowalczyk and coworkers[9, 10)] have made an experimental survey of $4\underline{s}$ and $4\underline{p}$ XPS spectra for the elements from $_{42}$Mo to $_{83}$Bi and a selection of those results are shown in Fig. 1. The data clearly demonstrate how a strong

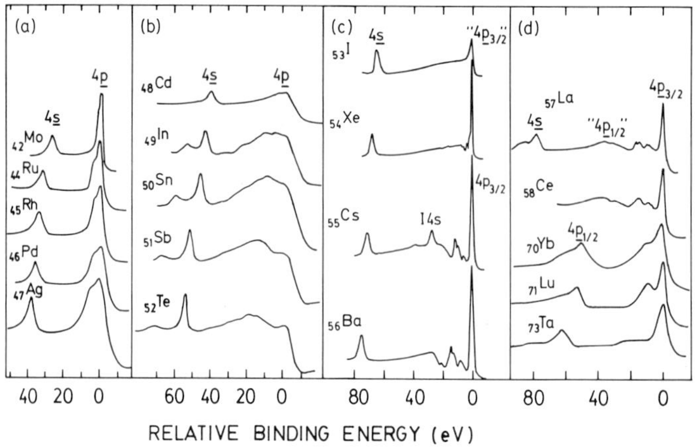

Fig. 1a–d. $4\underline{s}$, $4\underline{p}$ X-ray photoelectron spectra in the range $_{42}$Mo to $_{73}$Ta ((**a**), (**b**) and (**d**) from[9, 10], (**c**) from[7, 8])

interaction mechanism, in particular in the case of the $4\underline{p}$ peak, builds up at low Z and gradually loses its importance at high Z. The nature of this interaction mechanism is that the $4\underline{p}$ core hole can fluctuate to intermediate levels with two $4\underline{d}$ holes and an excited bound or continuum f-electron and back again

$$4\underline{p} \leftrightharpoons 4\underline{d}^2 mf \tag{1}$$

(hole levels are underlined; m represents both bound and continuum levels) and as a consequence the $4\underline{p}$ core hole level becomes shifted and broadened. This is then a particular case of the well known general problem of a discrete level interacting with a continuum[11, 12], namely a core hole autoionizing via Auger transitions. Normally, the Auger process is a quite weak perturbation so that the transition rate can be calculated by the Golden Rule. However, if the interaction is very strong, the same process will also distort the electronic charge density surrounding the core hole, thereby shifting the core level energy and changing the decay rate. This is what seems to be happening in the sequence of XPS spectra shown in Fig. 1. The core holes involved in the interaction process in Eq. (1) all belong to the same main (n = 4) shell and the decay rate can become very large. McGuire[13] called this process a *super Coster-Kronig transition* but actually we need an even stronger name. Equation (1) can be viewed as the $4\underline{p}$ core hole making a Coster-Kronig transition to $4\underline{d}$ while creating a $4\underline{d}$ mf electron-hole excitation in the 4d-shell. However, for the elements around Xe the response of the 4d-shell is highly collective because the final state electron tends to be resonantly localized in compact 4f-like states[14]. Therefore, in a sense all of the orbitals involved in the transition (Eq. (1)) belong to the same main (n = 4) shell and in the following this collective super Coster-Kronig fluctuation and decay process will be referred to as a *giant Coster-Kronig (gCK) process.*

The development of the $4\underline{p}$ XPS spectra in Fig. 1 can be described with the help of Fig. 2 as follows: In $_{42}$Mo (Fig. 1a) the $4\underline{p}$ levels lie high above the $4\underline{d}^2$ continuum threshold but since the $4d \rightarrow f$ transitions have a strongly delayed onset, the continuum

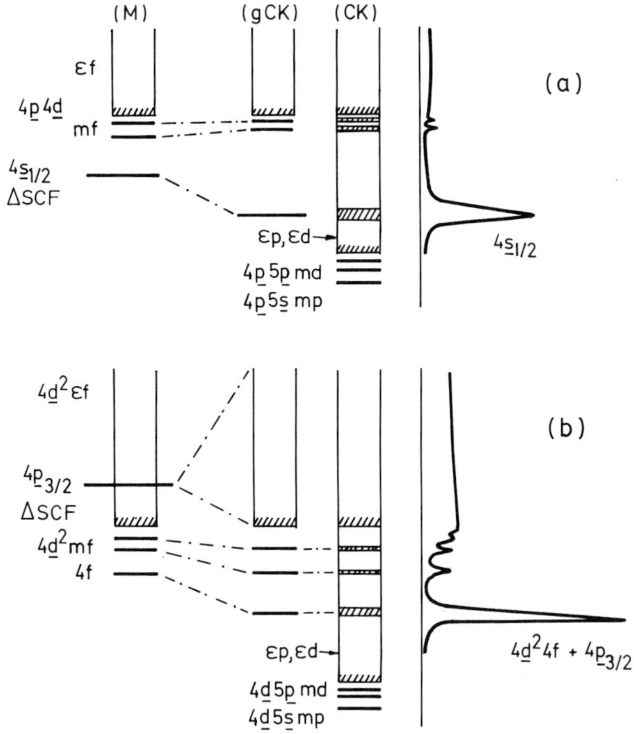

Fig. 2a, b. Schematic pictures of the energy level diagrams (total energies) for the interacting $4\underline{s}$ and $4\underline{p}$ core levels for the elements around $_{54}$Xe. (M) denotes monopole relaxation, (gCK) giant Coster-Kronig processes and (CK) Coster-Kronig processes. The figure illustrates how the levels shift and broaden when the interaction is turned on for **(a)** the $4\underline{s}_{1/2}$ level and **(b)** the $4\underline{p}_{3/2}$ level. The resulting core hole spectra have been drawn to illustrate the principle, rather than to be true replicas of experimental spectra

strength actually peaks at higher energies than the $4\underline{p}$ levels. The resulting interaction strength will be small and the $4\underline{p}$ core levels will look quite normal. As one proceeds to higher Z, the $4\underline{d}^2$ threshold will approach the $4\underline{p}$ levels from below while the peak in the $4\underline{d}^2$ εf continuum distribution will approach from above and at the same time become more contracted and peaked. As a result, the $4\underline{p}$ core levels will be pushed to lower binding energies relative to the ΔSCF positions and also become strongly broadened ($_{46}$Pd, $_{47}$Ag) until the core level structure has become virtually wiped out and the $4\underline{p}$ holes ceased to exist as stable excitations (Fig. 1b; $_{49}$In–$_{52}$Te). As the $4\underline{d}^2$ threshold comes close to the $4\underline{p}$ levels in $_{53}$I and $_{54}$Xe (Fig. 1c), discrete structure ($4\underline{d}^2$ 4f-like) becomes split off from the threshold (Fig. 2b). However, not until $_{55}$Cs does a proper $4\underline{p}_{3/2}$ line emerge, shifted by ~12 eV from the ΔSCF position, while the $4\underline{p}_{1/2}$ level basically remains wiped out. This situation seems to persist also in $_{56}$Ba, $_{57}$La and $_{58}$Ce and probably all the way through the rare earth metals. In the rare earth metals, however, multiplet structures, due to the open 4f-shell, seriously distort the spectra. Finally, from $_{70}$Yb and onwards, a more normal situation with distinct $4\underline{p}_{1/2}$ and $4\underline{p}_{3/2}$ levels seems to be recovered.

So far, we have said nothing about the $4\underline{s}$ XPS spectrum. Judging from the experimental spectra in Fig. 1 one might think that the $4\underline{s}$ levels look rather normal. However, there is again a giant Coster-Kronig fluctuation and decay process involving the $4\underline{s}$ hole

$$4\underline{s} \leftrightarrows 4\underline{p}4\underline{d}mf \tag{2}$$

Now the core hole makes Coster-Kronig transitions between $4\underline{s}$ and $4\underline{p}$ and the 4 d-shell responds more or less collectively by $4\underline{d}$ mf electron-hole pair transitions. The interaction strength is of the same order as in the $4\underline{p}$ case and the shift from the ΔSCF position goes from ~7 eV in $_{46}$Pd to ~11 eV in $_{56}$Ba. The difference is that the $4\underline{s}$ level never becomes wiped out because it slips below the $4\underline{p}4\underline{d}$ continuum thresholds before the decay part of the giant Coster-Kronig process reaches its maximum. This happens in the range $_{49}$In to $_{51}$Sb (see Sect. 4 and 6) and is clearly signalled by pronounced structure in the $4\underline{s}$ level and by narrowing of the line when going to $_{52}$Te. From there on the $4\underline{s}$ line width is mainly due to Coster-Kronig transitions and the $4\underline{s}$ line will appear rather narrow.

In this introduction I have outlined the original problem and most probably its solution. This solution was first suggested by Lundqvist and Wendin[15] in order to interpret the $4\underline{p}$ XPS spectrum in Ba[7,8] and was subsequently extended by Wendin and Ohno[16–19] to the 4 s and $4\underline{p}$ XPS spectra in Xe and a few surrounding elements. The purpose of the present article is to give a physical description of the many-body theory of core level spectra and extend the applications to the range of elements shown in Fig. 1, as well as to other cases in atoms, molecules and solids. The discussion will be geared towards the physics of the mechanisms that shape core level spectra. For those readers who want more detailed information about the systems and the spectra and their relation to physics and chemistry, there are several original articles and review papers[7–10, 17–20].

2 The Photoionization Process

The problem of the static and dynamic behavior of a core hole is intimately connected with the question of *how* the core hole was created. The *energy level spectrum* of the core hole is uniquely given by the energy level spectrum of the residual ion but the *intensity distribution* is determined by the excitation process. In order to interpret a photoelectron spectrum we therefore have to understand the photoionization process which gave rise to the photoelectrons. The purpose of this section is to discuss in simple terms the physics of the photoionization process and to demonstrate how the various contributions to the photoelectron spectrum can arise. A more formal discussion is presented in Section 3.

2.1 Semi-Classical Picture of the Photoionization Process

In Fig. 3 is shown a real space, real time picture of the initial stage of the photoionization process. The effect of the photon is to create an electron-hole pair $(\underline{i}\,\varepsilon)$ and while the photoelectron is on its way out the surrounding electron cloud responds to the perturbation in a number of ways. The actual importance of these many-electron effects depends on the system, the core level and the photon energy. Later on, some of the systematics of these effects will be discussed. Their nature can easily be understood from Fig. 3 where the "flow pattern" of other electrons is indicated. When the photoelectron and the hole separate they form an electric dipole which grows with time. For small separations between the electron and the hole, the surrounding electrons will respond to the dipolar field and the effect will be to set up an *effective dipole moment*. This can also be viewed as *collective* screening of the external electric field of the photon and is described by the random phase approximation with exchange (RPAE)[14, 21–25]. As the separation of the electron-hole pair increases, the surrounding electrons will start to notice the individual charges of the hole and the photoelectron and respond in order to screen out these charges by avoiding the photoelectron and collecting around the hole. As is evident from Fig. 3, this *relaxation process* happens in response to the field of the *electron-hole pair*.

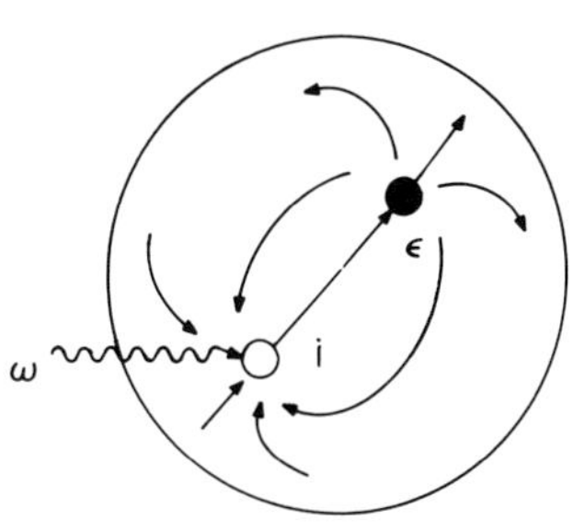

Fig. 3. Real space, classical picture of the photoionization process and the atomic response

One can therefore, in general, not separate the relaxation and shake up effects induced by the hole from the corresponding effects induced by the photoelectron. These considerations are particularly important in the photoionization threshold region where the speed of the photoelectron is relatively low.

2.2 The One-Electron Picture

In a real system any given elctron moves in the instantaneous potential of all the other electrons and its motion is therefore correlated with the motion of the other electrons. If instead we consider the *average* potential we obtain the Hartree-Fock one-electron picture where any given electron only interacts with the static average charge distribution of the other electrons. In the follwing we shall work in a HF zero-order basis and the energy eigenvalues will be denoted by E_i^0. In a strict one-electron picture an electron feels the same potential whether it is in the ground state or in an excited state. This means that all final state interactions are neglected and in the one-electron photoionization process there is no response in the form of polarization, relaxation, shake up etc. (see Fig. 4). The one-electron photoionization cross section then simply becomes (expressed in atomic units; see e.g.[24, 25])

$$\frac{d\sigma(\omega)}{d\varepsilon} = 4\pi^2 \alpha a_0^2\, \omega \sum_i |\langle i|z|\varepsilon\rangle|^2 \delta(\varepsilon - E_i^0 - \omega) \tag{3}$$

where $z = r \times \cos\theta$ is the dipole operator, ε the kinetic energy of the photoelectron, α the fine structure constant (= 1/137) and a_0 the Bohr radius (= 0.529×10^{-8}). The corresponding photoabsorption cross section is obtained by integrating over the photoelectron energy

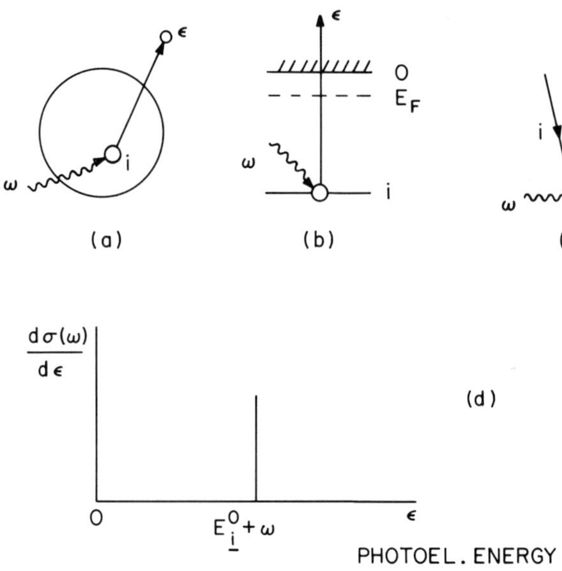

Fig. 4a–d. One-electron pictures of photoionization: **(a)** "Real space" picture, **(b)** one-electron level picture, **(c)** lowest-order Feynman diagram and **(d)** photoelectrum spectrum

$$\sigma(\omega) = 4\pi^2 \alpha a_0^2\, \omega \sum_{i} \int d\varepsilon |\langle i|z|\varepsilon\rangle|^2 \delta(\varepsilon - E_i^0 - \omega) \tag{4}$$

In the one-electron picture the photoelectron spectrum thus consists of a number of sharp peaks (Fig. 4 d), one for each core level, with the binding energy given by the negative of the HF eigenvalue (Koopmans' theorem[26]).

The δ-function in Eq. (3) contains information about the energy level structure and represents the simplest example of a *spectral function* for a core hole. By defining the spectral function $A_i\ (\varepsilon - \omega)$ as

$$A_i(\varepsilon-\omega) = \delta(\varepsilon - E_i^0 - \omega) \tag{5}$$

the photoelectron spectrum in Eq. (3) can be written as

$$\frac{d\sigma(\omega)}{d\varepsilon} = 4\pi^2 \alpha a_0^2\, \omega \sum_{i} |\langle i|z|\varepsilon\rangle|^2\, A_i(\varepsilon-\omega) \tag{6}$$

This result is actually valid far beyond the one-electron approximation because in the presence of interaction between the core hole and the surrounding electron cloud the spectral function $A_i(\varepsilon-\omega)$ will describe the full core level spectrum with shifted and broadened main lines, satellite lines and continua.

For a deep core hole in the XPS regime it is quite reasonable to explain the breakdown of the one-electron approximation in terms of interaction between the core hole and the average electronic medium. However, for core holes in the outer shells, Fermi sea correlations can give significant contributions to satellite spectra and for low photoelectron energies interactions between the photoelectron and the residual ion can become important. In the rest of this section several of these aspects will be illustrated.

2.3 Fermi Sea Correlation Effects

In the HF picture, any given electron in the ground state only feels the potential from the average ground state charge distribution. Due to the Pauli exclusion principle an electron with a given spin component will tend to keep electrons with the same spin away and be surrounded by a so-called *Fermi hole (exchange hole)*. This effect is included in the HF picture but what is neglected is that there is also a *correlation hole*[23]: Due to electron-electron scattering in the ground state the electrons tend to avoid each other and, as a result, the electron density in the neighbourhood of any given electron will deviate from the HF density. This will cause the effective one-electron wave function to deviate from HF and consequently contain components of *unoccupied* HF orbitals (ground-state configuration mixing). The scattering process is shown in a number of different way in Fig. 5. The interaction causes electrons j and k to *fluctuate* to other levels (virtual excitation process). This results in a contribution to the correlation energy, the lowest order Feynman diagram of which is shown in Fig. 5 c. The physical meaning of the diagram is simply that two HF electrons in the ground state collide and their density distributions, i.e. their orbits, become disturbed. These disturbances interact via the Coulomb interaction and contribute to the correlation energy. The manner in which the electrons try to avoid each

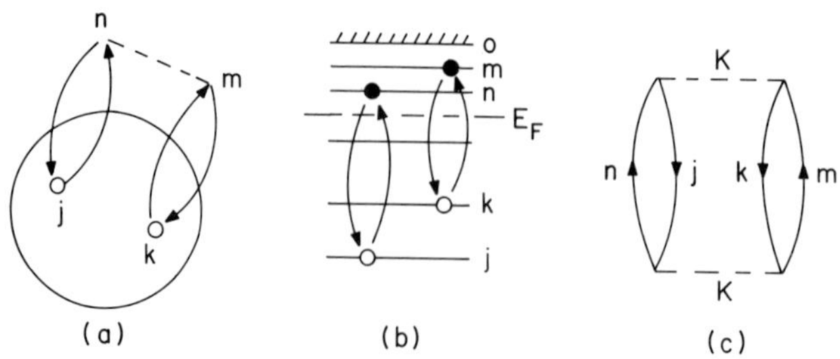

Fig. 5 a–c. Fermi sea correlation effects pictured in **(a)** "real space", **(b)** one-electron level scheme and **(c)** Feynman diagram

other depends on the angular momentum transfer K in Fig. 5c. For K = 0 the HF orbitals j and k become perturbed by excited states of the same symmetry (monopole density correlation). This leads to radial deformations and in/out correlation: *When the electron j is close to the nucleus electron k wants to be far away from the nucleus* and vice versa. For K = 1 the HF orbitals j and k become deformed in a dipolar manner (dipolar density correlation; cf. van der Waal's interaction) and this leads to an angular correlation: *The electrons prefer to be on opposite sides of the nucleus.* K = 2 leads to quadrupole density correlations and so on towards more and more complicated correlation patterns. Due to the correlation processes a photon can find a core electron virtually excited to states above the Fermi level (Fig. 6a). The consequence of ionizing a correlated electron is then a probability of finding the residual ion in an excited state, and as a result the core level will have a *spectrum of satellites* already due to ground-state correlations. Taking the 6s photoionization of Ba as an example, the 6s electrons may have a radial (in-out) correlation (Fig. 6b) leading to $6\underline{s}^2 7s$ and higher satellites (Fig. 6e) (a bar under an orbital index refers to a hole); they may have an angular (dipole) correlation (Fig. 6c) leading to $6\underline{s}^2 6p$ and higher satellites (Fig. 6e); as a final example, they may have angular (quadrupole) correlation (Fig. 6d) leading to $6\underline{s}^2 5d$ and higher satellites (Fig. 6e).

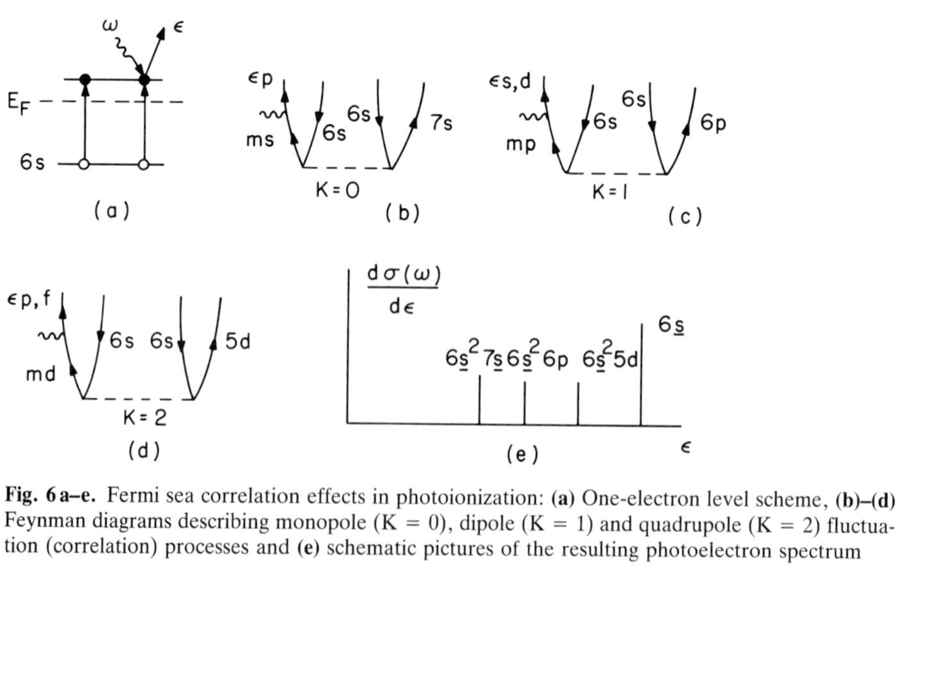

Fig. 6 a–e. Fermi sea correlation effects in photoionization: **(a)** One-electron level scheme, **(b)–(d)** Feynman diagrams describing monopole (K = 0), dipole (K = 1) and quadrupole (K = 2) fluctuation (correlation) processes and **(e)** schematic pictures of the resulting photoelectron spectrum

Regarding the importance of Fermi sea correlation effects for *satellite intensities*, on general grounds one would expect them to be relatively important for the outer shells and quite negligible for deep shells. For the outermost shells the excitation energy of two electron-hole pairs is as small as it can be and there may be prominent low-lying unoccupied levels for the pairs to be excited to. This is a situation when ground-state configuration mixing becomes important.

2.4 Shake-Up Processes

When an electron leaves the core region during a photoionization process, the surrounding electron cloud will contract in response to the change in the potential in order to screen out the positive charge of the core hole (Fig. 7 a) and this results in a relaxation shift of the core level energy. However, the contraction process also gives rise to "shock waves" in the electron cloud (cf. phonon and plasmon emission in solids[23)]) leading to excitation or ionization of the residual ion, so-called *shake-up* and *shake-off* processes (Figs. 7 b, d). Usually, atomic shells are very stable against radial perturbations and the total intensity of normal shake-up and shake-off satellite spectra is therefore rather weak, typically 10–15% of the intensity of the main core line in an XPS experiment where the photoelectron has high speed. In addition, if the photoelectron leaves the region of the core hole slowly, the surrounding electronic charge can relax adiabatically without creating shock waves and the shake-up satellite spectrum can become quenched. Diagrammatically, this is illustrated in Fig. 7 e where the photoelectron also interacts with the surrounding charge cloud and prevents it from seeing the full charge of the core hole. In the low energy region there is also the important possibility of the primary electron becoming "captured", giving off its excess energy to the shake-off electron (Figs. 7 c, e). This process is sometimes called conjugate shake-up and leads to a different satellite spectrum than the ordinary shake-up process.

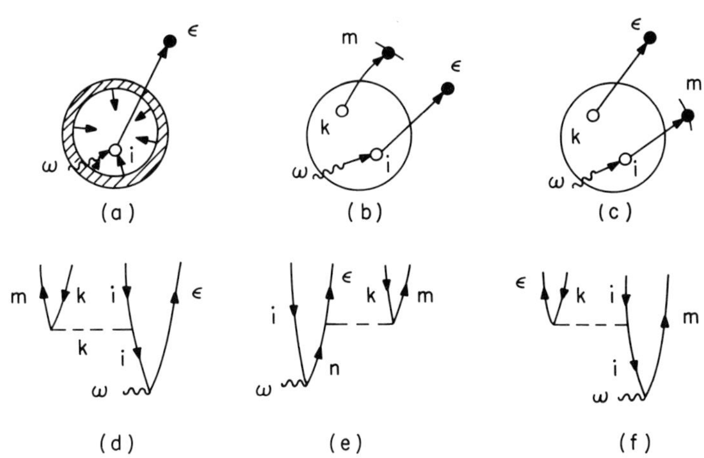

Fig. 7 a–f. Shake-up and shake-off processes in photoionization: **(a)** Real space picture illustrating contraction of the atomic charge density in response to the creation of a core hole; **(b, d)** Shake-up; **(c, f)** Conjugate shake-up and **(e)** inelastic scattering

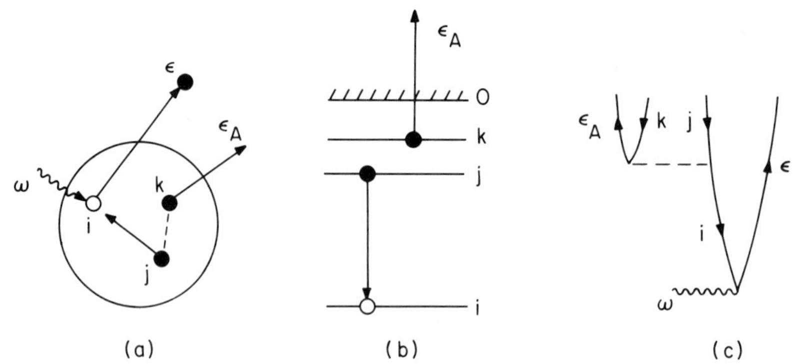

Fig. 8a–c. Photoemission of a primary electron with subsequent Auger emission

2.5 Auger Processes

In shake-up processes the core hole remains in the same orbital where it was created and the system merely responds to the change in the average charge. However, there is also the possibility that the core hole i makes a transition to another shell j as shown in Fig. 8, leading to the emission of an *Auger electron* from shell k. The core hole now only exists during a limited period of time and becomes spread in energy according to the Heisenberg uncertainty relation. If the interaction is weak, the perturbation of the ionic charge distribution can be neglected and the Golden Rule can be used for calculating the line width (transition rate) of the core hole. However, if the interaction is strong there is an appreciable relaxation which shifts the core level, distorts the charge distribution and modifies the transition rate. The Auger process must then be treated in a self-consistent manner. Such a formulation admits the breakdown of the quasi-particle picture and the one-electron orbital picture in the same way as the frequency of an overcritically damped oscillator loses its meaning. For very short times, there is some kind of excitation but there is no damped oscillatory long-time behaviour associated with any proper elementary excitation. This is the physics behind the missing core levels mentioned in the introduction and most of this article will be devoted to such problems.

3 Relaxation, Correlation and Decay of Core Holes

In this section we shall particularly study the dynamic properties of a core hole in terms of its self-energy and spectral function[15, 19, 23, 27–32]. This is a kind of model problem because one does not discuss by which physical mechanism the core hole is created. The hole is simply created in the system at a specific instant of time and destroyed at a later time. By studying the development of the core hole during this interval one gets a picture of how the core level strength becomes distributed over the various possible levels of the ionic system. Nevertheless, *since the creation of the core hole is sudden, the resulting spectral function is very closely connected to the X-ray photoelectron spectrum (XPS)* as already briefly discussed in Sect. 2.2, Eq. (6).

3.1 Self-Energy and Spectral Function for a Core Hole. The Quasi-Particle Picture

After the core hole has been introduced the system will start to relax and become shaken up, and eventually there may be an emission of Auger electrons. The key quantity for describing this behaviour is the core hole self-energy $\Sigma_{i}(E)$. This quantity tells how a core ough the reaction of the medium, giving rise to an effective core hole which not only describes the relaxed hole but also the shake-up spectrum and, as we shall see later, the Auger spectrum. The self-energy $\Sigma_{i}(E)$ includes everything that can happen to a core hole and, as the name suggests, it represents a generalized energy correction to the non-interacting core hole energy E_{i}^{0}. The *non-interacting core hole* is described by the Green's function

$$G_{i}^{0}(E) = \frac{1}{E - E_{i}^{0} - i\delta} \tag{7}$$

and from this one obtains the *spectral function*

$$A_{i}(E) = \frac{1}{\pi} \operatorname{Im} G_{i}^{0}(E) \tag{8a}$$

$$= \delta(E - E_{i}^{0}) \tag{8b}$$

With $E = \varepsilon - \omega$ we see that this is identical to the previous definition of the spectral function in Eq. (5). The non-interacting spectrum simply consists of a δ-function peak with unit strength at the position of the Hartree-Fock orbital eigenvalue (Koopmans' theorem). Switching on the interaction with the surrounding electron cloud the non-interacting or *bare* core hole propagator $G_i^0(E)$ becomes modified to an interacting or *dressed* core hole propagator (Green's function) $G_i(E)$, as shown in Figs. 9a, b. The infinite sequence of diagrams forms a geometrical series and the dressed core hole propagator simply becomes

$$G_i(E) = \frac{1}{E - E_i^0 - \Sigma_i(E)} \tag{9}$$

The spectral function for the *interacting core hole* is given by (cf. Eq. (8a))

$$\begin{aligned} A_i(E) &= \frac{1}{\pi} \operatorname{Im} G_i(E) \\ &= \frac{1}{\pi} \frac{\operatorname{Im} \Sigma_i(E)}{[E - E_i^0 - \operatorname{Re} \Sigma_i(E)]^2 + [\operatorname{Im} \Sigma_i(E)]^2} \end{aligned} \tag{10}$$

and the XPS photoelectron cross section by

$$\frac{d\sigma(\omega)}{d\varepsilon} = 4\pi^2 \alpha a_0^2\, \omega \sum_i |\langle i|z|E + \omega\rangle|^2 A_i(E) \tag{11}$$

Equations (9)–(11) have been derived on the assumption that the atomic *main* shells are well separated in energy and space. The rotational invariance forces a p-hole, say, to remain a p-hole, but in principle it can scatter to other single-hole p-states under the influence of various dynamic interactions. Therefore, the self-energy, the Green's function and the spectral function all become matrices $\Sigma_{ij}(E)$, $G_{ij}(E)$ and $A_{ij}(E)$ with off-diagonal elements. However, since the hole must hop between *different main shells*, this represents a very small perturbation, and for normal atomic systems the off-diagonal part of the self-energy is usually neglected. One can imagine this approximation to be less good or even to break down for systems with densely spaced one-electron orbitals, e.g. in the outer valence region of many molecules.

The structure of the spectral function (photoelectron spectrum) obviously depends on the structure of the self-energy $\Sigma_i(E)$, i.e. on the structure of the *ionic excitation spectrum*. Fig. 9c describes how the core hole i interacts with the average ground-state charge density and Fig. 9d describes the corresponding static exchange process. Working in the Hartree-Fock (HF) picture, these first-order self-energy diagrams are then included to all orders and do not appear any more. The lowest order contribution to the self-energy is then given by the second-order diagrams in Figs. 9e–g. The direct process in Fig. 9e gives the result

$$\Sigma_i(E) = \mathop{\mathrm{S}}_{mjk} \frac{|\langle jk|1/r_{12}|i\,m\rangle|^2}{E_m^0 - E_j^0 - E_k^0 + E - i\delta} \tag{12}$$

S denotes a summation over discrete levels and integration over the continuum. In this general discussion, we omit the contribution from the exchange process (Fig. 9f) which will only modify the matrix elements in Eq. (12), and we also omit the ground-state

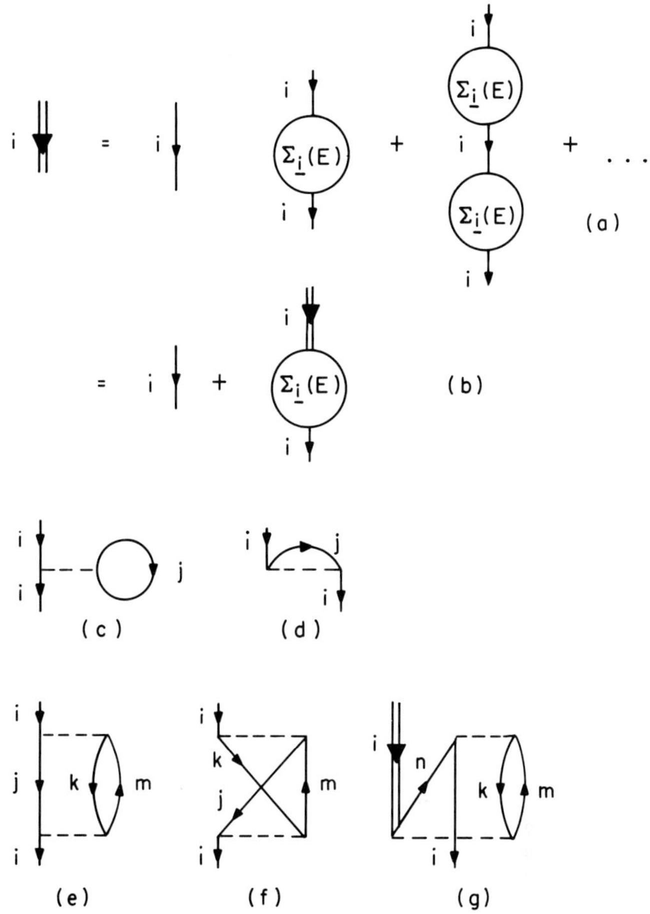

Fig. 9 a–g. The core hole self-energy $\Sigma_{\underline{i}}(E)$. **(a)** Expansion of the single-hole Green's function in terms of the self-energy $\Sigma_{\underline{i}}(E)$ which can be summed to give **(b)** the Dyson equation in diagrammatic form. **(c, d)** First-order contributions to $\Sigma_{\underline{i}}(E)$ including self-interaction, **(c)** the Hartree potential, **(d)** the Fock non-local exchange potential, **(e–g)** second-order contributions to the self-energy $\Sigma_{\underline{i}}(E)$: **(e)** direct (optical potential), **(f)** exchange and **(g)** Fermi sea correlation

(Fermi sea) correlation process in Fig. 9g. In this lowest order approximation the ionic excitation spectrum is given by a zeroth-order HF type of spectrum, which need not be a good approximation to the physical spectrum. However, since we are dealing with a normal system which does not show phase transitions, an exact description of the self-energy will only lead to shifts of energy levels and a redistribution of interaction strength. The analytical structure will therefore not change and as long as we only discuss the general character of the spectrum, the second-order self-energy expression in Eq. (12) is quite adequate.

In order to calculate the spectral function $A_{\underline{i}}(E)$ in Eq. (10) we need the real and imaginary parts of the self-energy $\Sigma_{\underline{i}}(E)$

$$\mathrm{Re}\,\Sigma_{\underline{i}}(E) = \sum\!\!\!\!\!\!\int_{mjk} \frac{|\langle jk|1/r_{12}|i\,m\rangle|^2}{E_m^0 - E_{\underline{j}}^0 - E_{\underline{k}}^0 + E} \qquad \text{(a)}$$

$$\mathrm{Im}\,\Sigma_{\underline{i}}(E) = \sum\!\!\!\!\!\!\int_{mjk} |\langle jk|1/r_{12}|i\,m\rangle|^2 \delta(E_m^0 - E_{\underline{j}}^0 - E_{\underline{k}}^0 + E) \qquad \text{(b)} \qquad (13)$$

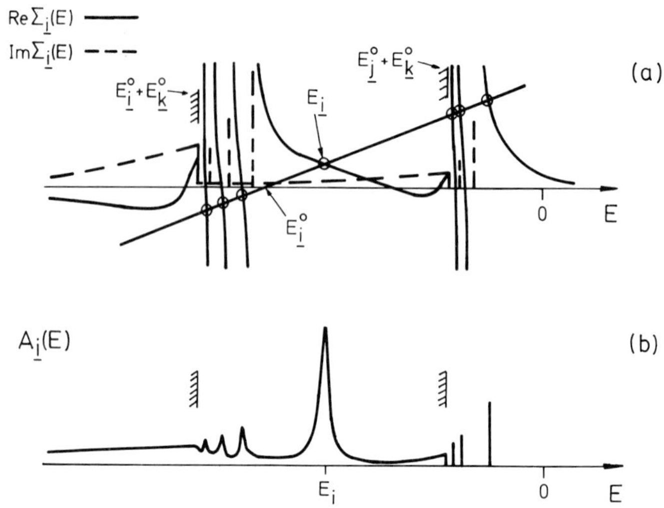

Fig. 10 a, b. Schematic, typical pictures of (**a**) the real and imaginary parts of the self-energy, $\mathrm{Re}\Sigma_{\underline{i}}(E)$ (—), and $\mathrm{Im}\Sigma_{\underline{i}}(E)$ (---), and a graphical solution of the Dyson equation, Eq. (15), and (**b**) the resulting spectral function $A_{\underline{i}}(E)$

$ denotes a principle part integration. The structure of $\mathrm{Re}\,\Sigma_{\underline{i}}(E)$ and $\mathrm{Im}\,\Sigma_{\underline{i}}(e)$ is shown in Fig. 10a in the typical situation that it describes a shake-up process ($\underline{j} = \underline{i}$ and an Auger process ($\underline{j} \neq \underline{i}$; $E^0_{\underline{j}} > E^0_{\underline{i}}$). $\mathrm{Im}\,\Sigma_{\underline{i}}(E)$ consists of a number of δ-function peaks at the discrete zeroth-order excitation energies of the ionic system and a continuum above the shake-off threshold $E^0_{\underline{i}} + E^0_{\underline{k}}$ and the Auger threshold $E^0_{\underline{j}} + E^0_{\underline{k}}$. $\mathrm{Re}\,\Sigma_{\underline{i}}(E)$ oscillates rapidly in regions where $\mathrm{Im}\,\Sigma_{\underline{i}}(E)$ is strongly peaked, i.e. always in the discrete region below a continuum threshold but often also in the very continuum above threshold. This is easily realized since the real and imaginary parts of the self-energy fulfil a Kramers-Kronig type of dispersion relation

$$\mathrm{Re}\,\Sigma_{\underline{i}}(E) = \frac{1}{\pi}\,\mathrm{PP}\int_{-\infty}^{\infty} \frac{\mathrm{Im}\,\Sigma_{\underline{i}}(E')}{E - E'}\,\partial E' \tag{14}$$

Evaluating the spectral function $A_{\underline{i}}(E)$ using the model spectrum for $\Sigma_{\underline{i}}(E)$ in Fig. 10a, we obtain a core level spectrum like that shown in Fig. 10b. The spectral function will have resonance maxima whenever

$$E - E^0_{\underline{i}} - \mathrm{Re}\,\Sigma_{\underline{i}}(E) = 0 \tag{15}$$

provided the corresponding value of $\mathrm{Im}\,\Sigma_{\underline{i}}(E)$ is not too large. The positions of the peaks correspond to the solutions of this so-called *Dyson equation* (Eq. (15)) and are given by the points of intersection between the straight line $E - E^0_{\underline{i}}$ and the curve for $\mathrm{Re}\,\Sigma_{\underline{i}}(E)$, as shown in Fig. 10a. If the spectral function contains a well-defined resonance peak at $E = E_r$, the strength and width of this peak can be found by Taylor expansion of the Dyson equation

$$E - E_i^0 - \operatorname{Re}\Sigma_i(E) \cong Z_i(E_r)^{-1}(E - E_r) \tag{16}$$

$$Z_i(E) = \left(1 - \frac{\partial}{\partial E}\operatorname{Re}\Sigma_i(E)\right)^{-1} \tag{17}$$

giving the well-known result of a *Lorentzian peak*

$$A_i(E) \cong \frac{Z_i(E_r)}{\pi}\frac{\Gamma_r/2}{(E - E_r)^2 + (\Gamma_r/2)^2} \tag{18}$$

with the full width at half maximum (FWHM) $\Gamma_r \equiv \Gamma(E_r)$

$$\begin{aligned} \Gamma(E) &= 2\,Z_i(E)\,\operatorname{Im}\Sigma_i(E) && \text{(a)} \\ \Gamma_r &\cong Z_i(E_r)\,2\pi|\langle jk|1/r_{12}|im\rangle|^2 \quad \text{at resonance.} && \text{(b)} \end{aligned} \tag{19}$$

The strength of the resonance peak is given by the slope of $\operatorname{Re}\Sigma_i(E)$ at $E = E_r$ (Eq. (17)). If the slope is zero, the so-called *renormalization factor* $Z_i(E)$ will be unity while if the slope is very large $Z_i(E)$ will tend to zero. It should be emphasized that Eq. (18) is only meaningful if the line shape parameters vary slowly over the width of the line. If there is strong dispersion, i.e. strong energy dependence of $\operatorname{Re}\Sigma_i(E)$ and $\operatorname{Im}\Sigma_i(E)$, the line will become asymmetric or even go over into a non-resonant continuum and one then has to go back to the full expression for the spectral function $A_i(E)$ in Eq. (10).

The particular structure of the self-energy and the spectral function shown in Fig. 10b with a shake-up process and an Auger process represents a normal situation. The shake-up excitations are usually separated by a considerable energy gap from the relaxed single, core-hole level $E = E_i$. The real part of the self-energy, $\operatorname{Re}\Sigma_i(E)$, will then vary rather slowly around $E = E_i$ and the peak strength will be rather close to unity, typically $Z_i(E) \sim 0.8$–0.9 for atomic shake-up processes. The strength stolen from the main peak will be transferred to the shake-up satellite lines, to the shake-off continuum and also to the satellites and the continuum associated with the Auger process. The effect of the interaction is to spread the unit strength of the non-intracting core hole δ-function peak over a wide energy range. The conservation of spectral strength can be expressed in terms of *sum rules* involving the spectral function[15, 23, 27–31)]

$$\int_{-\infty}^{\infty} A_i(E)\,dE = 1 \tag{20}$$

$$\int_{-\infty}^{\infty} E\,A_i(E)\,dE = E_i^0 \tag{21}$$

Equation (21) simply states that if the core hole is created suddenly, the HF-frozen core (Koopmans') energy E_i^0 is the centre-of-gravity of the resulting spectral distribution (provided Fermi sea correlations are neglected; this is usually a good approximation for deep holes). The normal situation is then that one single solution of Eq. (15), the single relaxed core hole, picks up most of the spectral strength, as suggested in Fig. 10b. The non-interacting core hole then primarily develops into a dressed, relaxed core hole with energy E_i when the interaction is turned on. This is the so-called *quasi-particle* picture of a core hole. With the help of Eqs. (21) and (22), the relaxation shift Δ_i of the hole can be written as (see also Manne and Åberg[33)])

$$\Delta_i \equiv E_i - E_i^0 = \int_{-\infty}^{\infty} (E_i - E)A_i(E)dE , \tag{22}$$

i.e. the moment of the satellite distribution with respect to the relaxed core hole position E_i balances the moment of the centre-of-gravity. The sign of the relaxation shift Δ_i obviously depends on which side of the main line the *centre-of-gravity of the satellite spectrum* is located. Normally, the monopole relaxation process (Sect. 3.2) dominates, the shake-up and shake-off satellites pushing the core hole strongly towards lower binding energy while the low-binding energy Auger-type satellites only very weakly push in the other direction. This will then lead to a normal healthy-looking core line which dominates the XPS spectrum. However, in recent years one has experimentally observed many cases where the main lines seem to be strongly perturbed or even missing. Typically, this situation can arise in two ways: Either (a) some kind of shake-up satellite levels move down to the region of the main line, or (b) part of the Auger-level structure happens to overlap the region of the main line. In both cases, there will be very strong dispersion in the region where the main line would be, i.e. $\mathrm{Re}\,\Sigma_i(E)$ will vary rapidly with energy. The quasi-particle strength $Z_i(E)$ can then become very small for the solution corresponding to a normally relaxed core hole and most of the spectral strength can become spread over a number of other excitations, e.g. a range of continuum energies. This represents, by definition, a breakdown of the quasi-particle picture and the concept of a well-defined core hole is then no longer valid. However, it is also possible that a considerable part of the spectral strength becomes concentrated to a particular excitation other than the simple core hole and then we have a quasi-particle picture of a more complicated kind.

3.2 Static and Dynamic Relaxation and Relation to Correlation and Symmetry Breaking

The response of an atomic, molecular or solid system to the creation of a core hole can, as always, be described in a real space, real time picture or, alternatively, in the corresponding Fourier space of energy and orbital representation. In a real space and time picture, which is more appealing to one's intuition, one can imagine a core hole as a wave packet of one-electron states forming a localized positive charge distribution and surrounded by a cloud of negative screening charge, as so-called *quasi-hole*. If the hole is stationary, the *screening is static* and one talks about static relaxation. If, instead, the charge distribution of the hole has a time dependence, e.g. a hole which oscillates between the nuclei in a molecule or moves along in the valence band of a semiconductor, the screening cloud will try to follow the motion of the hole. This then involves the *dynamic* response of the system and one talks about *dynamic screening* and *dynamic relaxation*.

By considering the bare hole as a localized charge distribution which breaks the symmetry of the system, the relaxation process leads to a correlation between the position of the hole and the position of the screening cloud. As a result, *the concepts of relaxation and correlation become inseparable*. The problem of symmetry breaking, correlation and collective excitations is well-known in the theory of many-particle systems[34–38)], and some aspects have recently been considered in applications to excitations of atoms and molecules[19, 39–42)].

We shall now analyze the self-energy $\Sigma_{\underline{i}}(E)$ of the hole in a way that connects the physical picture of the relaxation process with the more technical pictures based on expansions in terms of one-electron basis orbitals having the full symmetry of the system, involving concepts like monopole relaxation and shake-up, fluctuation and correlation. To this end, let us divide the self-energy $\Sigma_{\underline{i}}(E)$ into two parts corresponding to *relaxation (R)* (Figs. 9e, f) and *Fermi sea (ground-state) correlation (C)* (Fig. 9g)

$$\Sigma_{\underline{i}}(E) = \Sigma_{\underline{i}}^{R}(E) + \Sigma_{\underline{i}}^{C}(E) \tag{23}$$

and let us further divide the relaxation part into *static relaxation (SR)* and *dynamic relaxation (DR)*

$$\Sigma_{\underline{i}}^{R}(E) = \Sigma_{\underline{i}}^{SR}(E) + \Sigma_{\underline{i}}^{DR}(E) \tag{24}$$

Only the direct self-energy diagrams (e.g. Fig. 9e) can be interpreted in terms of classical physics. The exchange processes are sometimes important and must be included in actual calculations. However, they do not change the basic physical picture, and in this qualitative discussion we therefore only consider the direct processes.

The second-order relaxation diagram in Fig. 11a can be classified according to the character of the intermediate state one-electron orbitals j to which the initial hole in the one-electron orbital i jumps. Since there is a summation over all occupied levels j (cf. Eq. (12)), the diagram actually represents the construction of a wave packet for the hole in the intermediate state. The symmetry of the system is then formally broken, and the charge distribution of the relaxed hole can strongly deviate from the one-electron picture if the response of the system is favourable. Expressed in a different way, the self-energy in Fig. 11a describes a multipole expansion of the charge distribution of the hole, and if the dipole and higher terms are important, the hole cannot be thought of as existing in a one-electron symmetry orbital. Instead, in a snapshot picture, the angular distribution of the hole will appear as more or less *localized*.

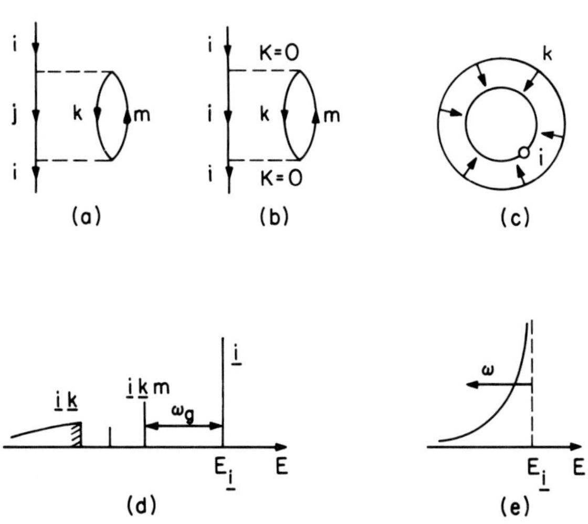

Fig. 11a–e. Monopole relaxation in response to a localized core hole: **(a)** general self-energy diagram; **(b)** monopole part of **(a)**; **(c)** "real space" picture of monopole relaxation, illustrating the radial contraction of a relaxing shell; typical core level spectra in **(d)** atoms and molecules and **(e)** metals

We can now define the connection between the physical relaxation processes and the internal structure of the direct self-energy diagram in Fig. 11 a:

Static relaxation (SR):	$E^0_{\underline{j}} = E^0_{\underline{i}}$	(25 a)
Static monopole relaxation (S0):	$E^0_{\underline{j}} = E^0_{\underline{i}},\ \underline{j} = \underline{i},\ K = 0$	(25 b)
Static higher-pole relaxation (SK):	$E^0_{\underline{j}} = E^0_{\underline{i}},\ \underline{j} \neq \underline{i},\ K = 1$	(25 c)
Dynamic relaxation (DR):	$E^0_{\underline{j}} \neq E^0_{\underline{i}}$ ($\underline{j}$ automatically $\neq \underline{i}$)	(26 a)
Dynamic Kth-pole relaxation (DK):	$E^0_{\underline{j}} \neq E^0_{\underline{i}},\ K$	(26 b)

In the remaining part of this section we shall discuss some of these processes in greater detail.

3.3 Static Monopole Relaxation and Shake-Up Processes

In a conventional picture of relaxation a core hole acts like a static, spherical charge distribution without any internal degrees of freedom and the associated electrostatic forces polarize the surrounding electron cloud. This gives rise to an interaction energy which shifts the position of the core level from the frozen HF value towards lower binding energy ([29, 43–46]and references therein). The classical monopole test charge character of the core hole is reflected in the self-energy $\Sigma_{\underline{i}}(E)$ (Fig. 11 b) in that the core hole line i goes continuously through the diagram without changing its state and in that there is no angular momentum transfer (K = 0) between the core hole and surrounding medium. The core hole $\underline{i}$ therefore perturbs orbital k by mixing in excited orbitals m with the same symmetry as k itself. The overall result is a radial (spherically symmetric) contraction of the ionic charge distributions (Fig. 11 c) and one talks about *static monopole relaxation* (Eq. (25 b)).

There is also the possibility of finding the ionic system in an *excited state* where the core hole i is accompanied by one or several other electron-hole pair excitations, the so-called *shake-up and shake-off excitations* ([44–46]and references therein). As long as the core hole does take up not angular momentum (K = 0), the ionic excitations must lead to higher total energies and the corresponding monopole shake-up and shake-off excitations must occur on the high-binding energy side of the main, relaxed, core line (Fig. 11 d). The static monopole relaxation shift $\Delta^{S0}_{\underline{i}}$ is obtained from the Dyson Equation (Eq. (10)) with the static part of the monopole self-energy

$$\Delta^{S0}_{\underline{i}} = E_{\underline{i}} - E^0_{\underline{i}} = \Sigma^{S0}_{\underline{i}}(E_{\underline{i}}) \tag{27}$$

and can be related to the static part of the atomic monopole polarizability[43]. $\Delta^{S0}_{\underline{i}}$ can be obtained from an integration over the monopole shake-up and shake-off spectrum (Eq. (22)). Finally, the static monopole relaxation shift can also (and perhaps most directly) be obtained through the ΔSCF method.

So far we have discussed the stationary state-level structure of the core hole. However, one can also consider the explicit time development of the core hole by looking at the time dependence of the response of the system ([47–50]and references therein). At time t = 0 the core hole is created in state i and the k-th shell, say, begins to relax in order to screen the perturbation. The inverse of the excitation energy ω_m of a certain shake-up excitation $\underline{k}m$ corresponds to a response time $\tau_m \sim \omega_m^{-1}$, characterizing that particular component of the screening charge. For very short times after the creation of the core hole only the high-energy ionic excitations have time to respond. This causes some relaxation of the ionic charge and the high energy part of the shake-off satellite spectrum is formed. The ionic system will continue to relax and to generate the shake-off and shake-up spectrum until $t \sim \tau_s \sim \omega_g^{-1}$, where the excitation energy gap ω_g is the lowest satellite excitation energy. Once this part of the charge has relaxed, the relaxation is complete and the satellite spectrum has reached its full strength.

3.4 Static Non-Monopole Relaxation, Shake-Up and Shake-Down

Normally, an atomic system is very stable against the perturbation of a static core hole. The large energy gap for monopole shake-up excitation limits the monopole polarizability and puts an effective limit on the radial charge contraction. However, the system need not necessarily be stable against *non-spherical* perturbations if it has an *open shell structure*. Core holes with angular momentum $l \geq 1$ have an additional degree of freedom in that they can transfer an angular momentum $K \geqslant 2$ to the surrounding electron cloud by jumping to different magnetic sublevels i′ (Fig. 12 a). Whether this will have any noticeable consequences depends on the energy-level structure of the residual ion. The presence of a core hole can make an empty, diffuse high-l orbital contract and collapse into the core (Fig. 12 b). The ionic system could then lower its energy by e.g. quadrupole distortion of the outer shell (Fig. 12 c). As a result, satellites could move down into the region of the main core level (Fig. 12 d) and pick up a large part of the core level strength. Off-hand, it is not obvious that the core level would be strongly perturbed because the coupling strength does not appear to be very large. However, experimental

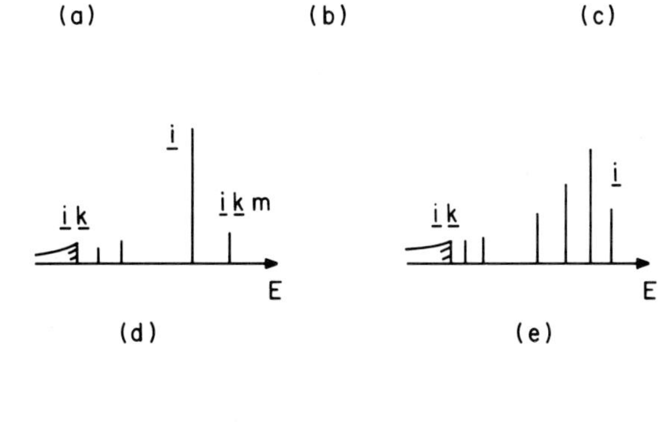

Fig. 12 a–e. Non-monopole (quadrupole) relaxation and shape distortion of a core hole in open shell type of situations: **(a)** Lowest order self-energy diagram; **(b)** one-electron picture of the pulling down of empty levels below the Fermi level; **(c)** shape distortion of one-electron orbitals due to quadrupole relaxation; **(d, e)** schematic core level spectra in the case of **(d)** closed and **(e)** open *ground state* shell structure

results[51–55] strongly suggest a breakdown of the quasi-particle approximation for the $5\,\bar{p}_{3/2}$ level in Ba[51–55] and the $3\,\bar{p}$ level in Ca[54, 55], and there are several other cases showing strong perturbations[54, 55]. Such behaviour might be quite common in atomic systems with pronounced outer shell s-d degeneracy as well as in molecules and solids. Cases of satellites with negative shake-up energy (Fig. 12 d), so-called *"shake-down"* satellites, are not uncommon, and we shall return to these interesting problems in later sections.

The above example is a special case where an open shell situation is *induced* by the spherical part of the core-hole electrostatic potential and where the subshells subsequently distort due to the non-spherical part of the interaction. In the more general case of an open shell atom, there is always a number of low-energy valence shell excitations which can be excited by the non-spherical part of the core-hole potential during the relaxation process. As a result, the distorted core hole can correlate with a number of modes of distortion of the open valence-shell charge distribution, and the core hole level is then split into several discrete sublevels. In a time-dependent picture, one might say that for very short times after its creation, the core hole does not see the open shell structure. Only for rather long times, comparable to the inverse of the multiplet energy splitting, does the open shell structure become important and the one-electron level is split into a main line and a number of often very prominent satellites, as indicated in Fig. 12 e.

A related but more extreme situation occurs in a metal because the conduction electrons are infinitely polarizable due to the absence of any excitation energy gap. The core hole will then continue to generate satellites with lower and lower excitation energy until all spectral strength lies in a satellite continuum and the main line has lost all of its strength (Fig. 11 e). Furthermore, in the absence of lifetime broadening the satellite continuum is singular as $\omega^{1-\alpha}$, the so-called *X-ray edge singularity*[48].

In a sense, one can say that the effect of the relaxation process is to broaden the spectral distribution of the core hole. This is sometimes called non-lifetime broadening because one is not dealing with any real decay process, i.e. lifetime broadening, and the main line is non-degenerate with the satellite spectrum. Only for very short times is the energy spread of the core hole large enough to make the main line and the satellites overlap, admitting intensity transfer between various parts of the spectrum. Experimentally, one can only detect the photoelectrons after a very long time when the relaxation process is complete. However, inside the system the core hole can be filled through an X-ray emission process or an Auger process *before* it has settled down to a stationary state. Effects of incomplete relaxation have recently been shown to exist for lattice relaxation in solids[56, 57] and in models for the relaxation of hole levels in molecules adsorbed on metal surfaces[50] (see also Sect. 8).

3.5 Dynamic Relaxation and Decay

In order to describe the correlation of motion between the core hole and the screening electrons, we must construct wave packets representing "individual" electrons in the atomic shells by superposing one-electron orbitals. In a perturbation expansion, electrons and holes will then appear to jump between different one-electron levels, and these processes can be discussed in terms of *fluctuation and decay*. Since the system is described as jumping between different configurations, the fluctuation process corresponds to con-

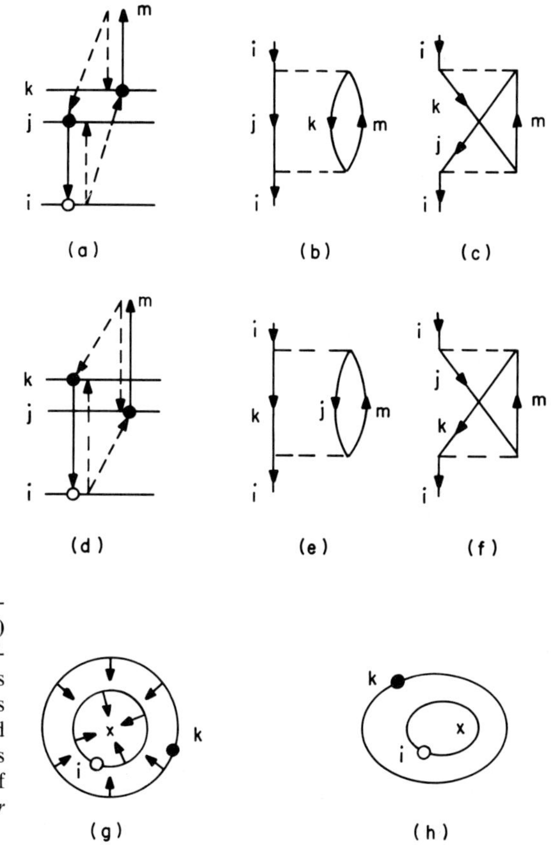

Fig. 13a–k. Dynamic relaxation and fluctuation processes of a hole. Direct **(b, e)** and exchange **(c, f)** processes for the self-energy. **(a, d)** One-electron level pictures of the fluctuation processes (full arrows denote decay of the hole, full plus dashed arrows denote fluctuation); **(g)** describes dynamic relaxation via *radial* distortions of the subshells while **(h)** describes *angular* distortions

figuration interaction (CI). However, CI as a computational method, is usually limited to *discrete* configurations, while here we are mainly interested in fluctuations involving a *continuum of states*.

The lowest order direct and exchange self-energy diagrams describing the fluctuations process are shown in Figs. 13b, c, e, f together with the corresponding one-electron level pictures in Figs. 13a, d. The effect of the fluctuation process is to hybridize the core orbital i by mixing in a spectrum of other orbitals j, thus breaking the symmetry restrictions on the one-electron orbitals. In a *time-independent* picture one would then work e.g. within an *unrestricted* HF scheme in terms of stationary one-electron orbitals which do not have the full symmetry of the system[34, 35, 37, 38, 40]. In a *time-dependent* picture one would instead describe the fluctuation process in terms of time-varying deformations of the one-electron orbital i and the corresponding dynamical response of the electronic charge cloud.

The correlation pattern depends on the angular momentum transfer K between the core hole and the screening charge. For $K = 0$ there is a correlation between the *radial* motion, *monopole fluctuation,* of the core hole and the screening charge (Fig. 13g) while for $K \geqslant 1$ there is an *angular* correlation. Fig. 13h describes the case of dynamic *dipolar fluctuation* ($K = 1$) where the core hole no longer is spherically distributed and where the screening charge tries to follow. In principle, there is also dynamic quadrupole (K = 2) and higher correlations.

It is interesting to note that in the quadrupole relaxation process discussed in Section 3.4 (Fig. 12) the core hole can be regarded as "fluctuating" between the different degenerate orbital magnetic sublevels. In this way, the core hole can form a static quadrupole moment and induce a quadrupole screening charge distribution.

Whenever *fluctuation* processes are discussed one also has to consider the complementary *dissipation* processes. If the core hole level $\underline{i}$ is degenerate with a continuum of intermediate states, then it can make a *real* transition to level $\underline{j}$ and excite an electron k to the continuum. In other words, the core hole can *decay* via an Auger process (Figs. 13 a, d, full arrows). Usually, the Auger fluctuation and decay process represents only a small perturbation so that the fluctuation level-shift can be neglected compared with the static monopole-relaxation shift (given by the ΔSCF method). However, in the case of the giant Coster-Kronig (gCK) fluctuation and decay process (Eqs. (1), (2)) mentioned in the introduction the whole correlation process essentially concerns only a single main shell so that the amplitudes of the distortions become very large and strongly correlated. As a result, the dynamic relaxation shift $\Delta_{\underline{i}}^{DR}$ can become a good deal larger than the static relaxation shift $\Delta_{\underline{i}}^{SR}$ (Sect. 3.3) and the combined effect of fluctuation and dissipation can be to smear the core level to a non-resonant continuum. The occurrence and detailed consequences of the gCK processes will be the subject of most of the remaining part of this review.

3.6 Relaxation and Correlation of two Core Holes

In our discussion of the dynamics of a single core hole we have so far used a simple one-electron model for the energy levels of the excited ionic state and also for the transition matrix elements leading to these states (Eq. (12)).

Although this lowest order approximation is enough for demonstrating the basic properties of the self-energy, it is inadequate for describing the details of the shake-up and shake-off spectrum and the structure deriving from Auger processes. Actually, in cases where the main core level overlaps with the structure in the self-energy, even the *qualitative* behaviour of the spectral function can be wrongly predicted. The obvious thing to be noticed is that the intermediate states in the self-energy also have to be eigenstates of the system. *The two core-hole states therefore must be allowed to relax* and, in principle, one should allow for shake-up effects and for Auger and autoionization processes leading to lifetime broadening of the excited levels of the ion, as well as for Fermi sea correlation effects. Therefore, in order to understand the dynamics of a single core hole one must also understand the dynamics of a double vacancy, and so on. This can be done within the general framework used for dressing up a single core hole, as shown in Fig. 14. Fig. 14 a describes how each of the core holes $\underline{j}$ and $\underline{k}$ relax independently of another and become dressed up by self-energies $\Sigma_{\underline{j}}(E)$ and $\Sigma_{\underline{k}}(E)$ describing relaxation and Fermi sea correlation. Furthermore, the two core holes will repel each other with an effective interaction $I_{\underline{jk}}(E)$ which describes the screened Coulomb repulsion plus non-linear effects (Fig. 14 b), so that the energy of the interacting core holes becomes

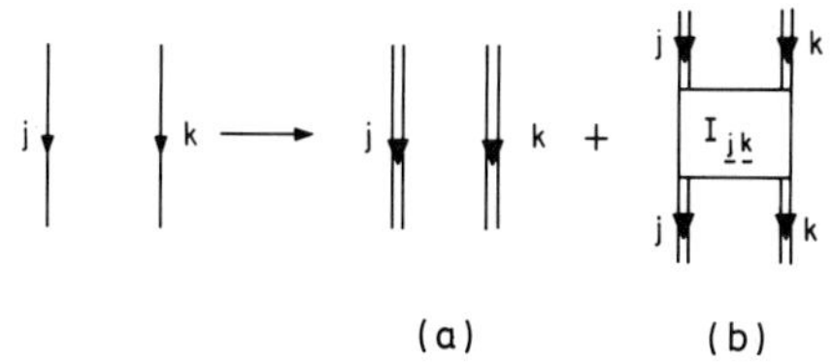

Fig. 14 a, b. Renormalization of a double hole level in terms of **(a)** independent, interacting (dressed) holes and **(b)** effective interaction between dressed holes

$$E_{\underline{j}}^{0} + E_{\underline{k}}^{0} \rightarrow E_{\underline{jk}}(E) = E_{\underline{j}}(E) + E_{\underline{k}}(E) - I_{\underline{jk}}(E) \tag{28}$$

In the case of two equivalent core holes and static monopole screening, the centre-of-gravity energy approximately becomes[18, 19]

$$E_{\underline{j}}^{2} = 2\,(E_{\underline{j}}^{0} + \Delta_{\underline{j}}^{S0}) - (F^{0}(j;j) - 2\,\Delta_{\underline{j}}^{S0}) \tag{29}$$

where $\Delta_{\underline{j}}^{S0}$ is the static monopole relaxation shift of a single core hole $\underline{j}$ and $F^{0}(j;j)$ is an ordinary Slater Coulomb integral. A similar result has been derived by Shirley[58] using the technique of Hedin and Johansson[43].

4 Survey of the Breakdown of the Quasi-Particle Picture in Core Level Spectra of Atoms

As mentioned previously, really strong many-electron effects and breakdown of the quasi-particle picture occur when (a) single and double core hole levels are nearly degenerate and simultaneously (b) the coupling strength is large. Both of these conditions are frequently fulfilled in systems with strongly polarizable shells, showing strong collective behaviour in e.g. photoionization[14, 24, 25]. These effects occur for systems where a *main shell is open* (although several *subshells* may be closed) *so that transitions can take place within that main shell.* As a rule of thumb, the strongest effects can be expected for the rare gas atoms from Ar and upwards and for a broad range of elements around them, for free atoms as well as in molecular and solid (metallic) form. For such systems with open main shells, the Z-dependence of single and double core-hole energies shows interesting trends. When a subshell starts to fill, it can often be ionized several times before single ionization of an inner subshell becomes favourable, i.e. the double vacancy has lower binding energy than the deeper single vacancy. However, the energy of a double vacancy state will rise more rapidly with atomic number Z than the energy of a single vacancy in the same main shell and therefore a level crossing will finally take place. A quasi-degeneracy may persist for certain level crossings while the main shell is being filled. However, when this process is completed there will be a rapid separation of the single and double hole levels. A concentrated overview of the most important cases of level crossing and strong perturbation and breakdown of the quasi-particle picture for core holes in atoms is given in Table 1.

The general behaviour outlined above is best illustrated by a few specific examples. In Fig. 15 are shown all of the single core-hole relativistic ΔSCF binding energies from $4\underline{s}_{1/2}$ to $5\underline{p}_{3/2}$ and most of the double core-hole binding energies from $4\underline{p}4\underline{d}$ to $5\underline{p}^2$ for the range of elements where the $4\underline{p}$ core level disappears (Fig. 1). From Fig. 15 the general Z-dependence of the core-level spectra can easily be understood. At the Cd end both the $4\underline{s}$ and $4\underline{p}$ levels are located in the (mainly f-) continua above the $4\underline{d}^2$ and $4\underline{p}4\underline{d}$ thresholds, respectively, and are consequently strongly broadened and shifted. Judging from the experimental XPS spectra (Fig. 1) the quasi-particle picture for a $4\underline{p}$ core hole might very well have broken down by Cd. In Fig. 15, as one proceeds towards higher Z, the $4\underline{s}_{1/2}$ level becomes degenerate with the $4\underline{p}4\underline{d}$ threshold structure around Sn, and for a number of elements around Sn one should observe structure in the $4\underline{s}_{1/2}$ peak region. This could be the explanation for at least part of the structure observed experimentally from In to Te in Fig. 1.

In what concerns the $4\underline{p}_{1/2}$ and $4\underline{p}_{3/2}$ levels, the coupling to the $4\underline{d}^2\varepsilon f$ continuum will increase with increasing Z and the core levels will be entirely wiped out. As the level crossing is approached there will be some discrete structure growing up below the $4\underline{d}^2$ threshold for I and Xe (Fig. 1) and probably not until Cs does the $4\underline{p}_{3/2}$ level emerge as a well-defined quasi-particle excitation. An interesting thing is that the $4\underline{d}^2$ level seems to

Table 1. Overview of cases with strong perturbations and breakdown of the one-electron picture for core holes in free atoms

	Process		Elements
I. Giant Coster-Kronig (gCK)-intra-main shell fluctuations			
(a)	$n\underline{s} \leftrightarrows n\underline{p}^2 nd,\ \varepsilon d$	$n = 3, 4, 5, 6, 7$	
		$n = 3$	Ar and surr. elements
		$n = 4$	Kr
		$n = 5$	Xe
		$n = 6$	Rn
		$n = 7$	$Z = 120$, Unknownium (ekaradium)
(b)	$n\underline{s} \leftrightarrows n\underline{p}\, n\underline{d}\, nf,\ \varepsilon f$	$n = 4, 5, 6$	
		$n = 4$	Sn
		$n = 5$	Ra
		$n = 6$	
(c)	$n\underline{s} \leftrightarrows n\underline{p}\, n\underline{f}\, ng,\ \varepsilon g$	$n = 5$	
(d)	$n\underline{p} \leftrightarrows n\underline{d}^2 nf,\ \varepsilon f$	$n = 4, 5, 6$	
		$n = 4$	Xe
		$n = 5$	Rn
		$n = 6$	$Z = 120$
(e)	$n\underline{p} \leftrightarrows n\underline{d}\, n\underline{f}\, ng,\ \varepsilon g$	$n = 5$	$Z = 120$
(f)	$n\underline{d} \leftrightarrows n\underline{f}^2 ng,\ \varepsilon g$	$n = 5$	$Z = 120$
II. Super Coster-Kronig (sCK) fluctuations – all *hole* levels in the same main shell			
(a)	$n\underline{s} \leftrightarrows n\underline{p}\, n\underline{d}\, mf,\ \varepsilon f$	$n = 3,\ m = 4$	Cu-Kr ($25 < Z < 40$)
		$n = 4,\ m = 5$	
(b)	$n\underline{p} \leftrightarrows n\underline{d}^2 mf,\ \varepsilon f$	$n = 3,\ m = 4$	Cu-Kr ($25 < Z < 40$)
		$n = 4,\ m = 5$	4 f-elements and upwards
(c)	$n\underline{d} \leftrightarrows n\underline{f}^2 mg,\ \varepsilon g$	$n = 4,\ m = 5$	
III. Coster-Kronig (CK) fluctuations			
(a)	$n\underline{s} \leftrightarrows n\underline{p}\, m\underline{d}\, m'f,\ \varepsilon f$	$m', m > n,\ m' \geqslant m$	

exhibit a similar Z-dependence as the $4\,\underline{p}_{1/2}$ and $4\,\underline{p}_{3/2}$ levels throughout the filling of the 4 f-subshell (Fig. 15) and the near degeneracy of the $4\,\underline{d}^2$ and $4\,\underline{p}_{1/2}$ levels therefore persists through the rare-earth series. After that, the levels rapidly separate and the $4\,\underline{p}_{1/2}$ level also emerges as a well-defined quasi-particle excitation (Fig. 1).

In Fig. 15 several other interesting trends can be seen. The $5\,\underline{s}_{1/2}$ level, for instance, lies very close to the $5\,\underline{p}^2$ threshold and will couple strongly to the $5\,\underline{p}^2$ md, εd excitation channel. It will therefore be strongly shifted by giant Coster-Kronig fluctuations and there will be prominent satellite lines. However, in contrast to the case of the $4\,\underline{s}$ and $4\,\underline{p}$ levels, there is no level crossing, the $5\,\underline{s}_{1/2}$ level always lying below the $5\,\underline{p}^2$ threshold and remaining a well-defined quasi-particle excitation. The giant Coster-Kronig fluctuation process for the $5\,\underline{s}$ level has analogues in all the rare gases and surrounding elements from $_{18}$Ar and upwards.

Figure 15 also shows a case of level crossing involving the weaker *Auger* process, namely $4\,\underline{d} \rightarrow 5\,\underline{s}\ 5\,\underline{p}$ ml Auger fluctuation and the decay of a $4\,\underline{d}$-core hole in $_{49}$In.

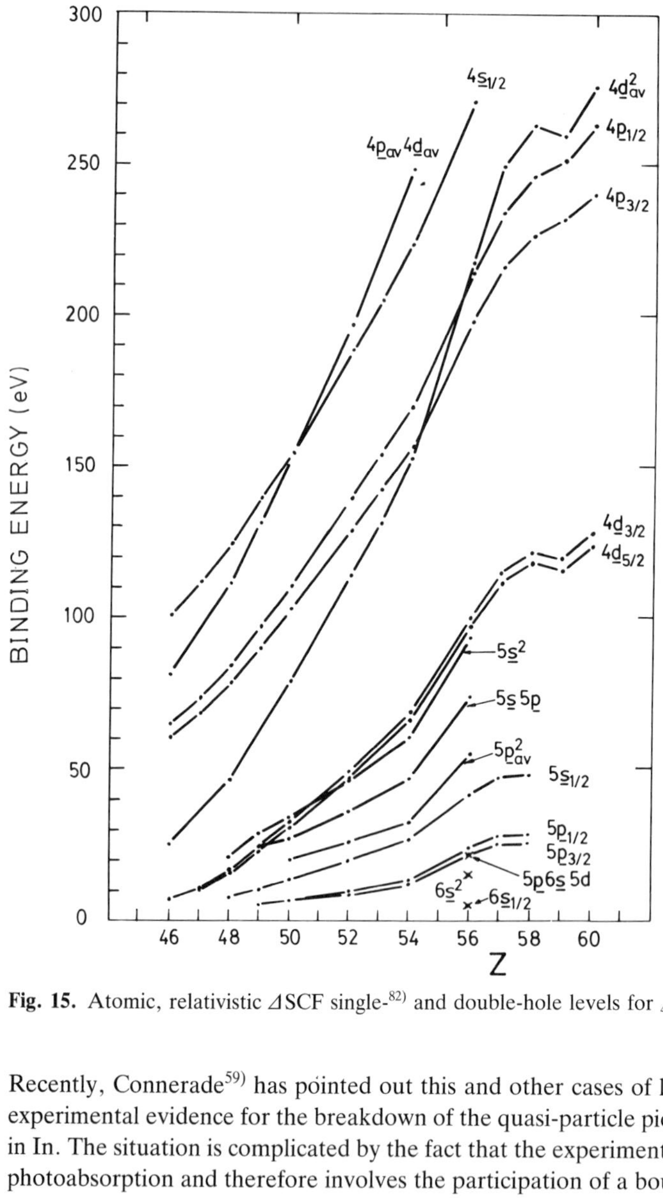

Fig. 15. Atomic, relativistic ΔSCF single-[82] and double-hole levels for $_{46}$Pd to $_{60}$Nd

Recently, Connerade[59] has pointed out this and other cases of level crossing and given experimental evidence for the breakdown of the quasi-particle picture for a 4 d-core hole in In. The situation is complicated by the fact that the experiment concerns 4 d-threshold photoabsorption and therefore involves the participation of a bound or very slow photoelectron which perturbs the picture of an interacting core hole. One therefore has to consider interactions between discrete nd continuous single and double excitations.

Another case of level crossing involves the 5 p levels and occurs from $_{55}$Cs, $_{56}$Ba and upwards (Fig. 15). This leads to extremely complicated photoelectron[53, 60], Auger electron[53–55] and photoabsorption spectra[51, 52] (for further details see Sect. 6.4).

The behaviour of the 4 s, 4 p core levels around Xe is based on the existence of *four subshells,* basically three (4 s, 4 p, 4 d) filled and one (4 f) empty. In order to see further examples of complete breakdown of the one-electron picture one has to go to higher Z, to the n = 5 shell in the 5 f group and preceding elements (Fig. 16; Table 1, entry I). In

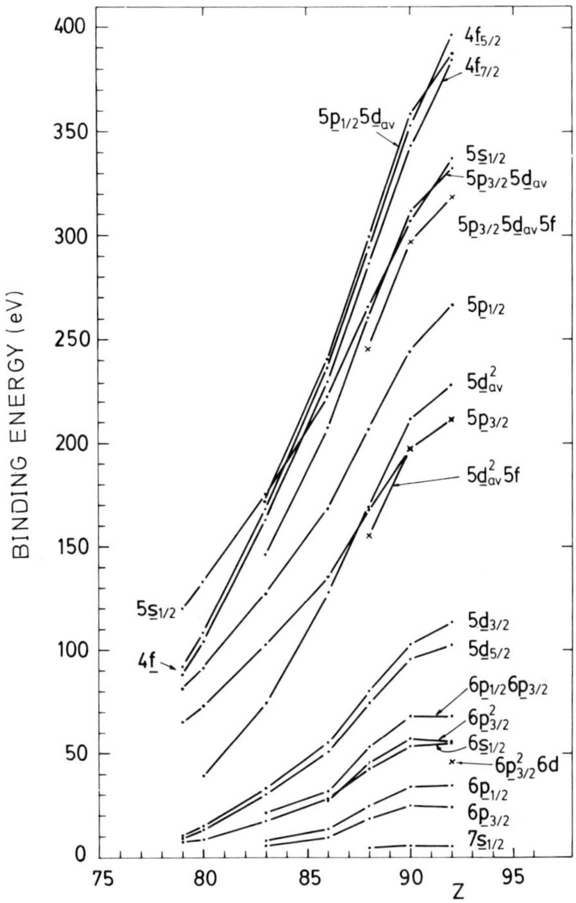

Fig. 16. Atomic, relativistic ΔSCF single-[82] and double-hole levels for ${}_{79}$Au to ${}_{92}$U

the case of a 5p hole there is now strong experimental and theoretical evidence for such behaviour in the range $80 < Z < 100$[10, 61–65], and in Section 6 also this and other cases will be incorporated into the general giant Coster-Kronig framework.

Unfortunately, the physical (or at least known) part of the periodic system ends just before several new interesting giant Coster-Kronig interaction processes have the possibility to become important, in the region of the 5f/6d group elements and beyond. A number of degeneracies will develop in the region of superheavy elements. At very high Z, the transition periods will become very long and contain a large number of open shells, perhaps resulting in an electron-liquid like behaviour of the outer region and in prominent satellite structures on most core levels. However, from an electronic point of view, the elements in the range $Z = 95 - 120$ behave in a normal way. The elements $Z = 118$ and $Z = 120$ correspond to a rare gas and an alkaline earth, respectively. Apart from strong many-electron effects in the 7s and 6p levels, there is now for the first time the possibility of having important effects of giant Coster-Kronig fluctuation and decay processes also in the 5d levels (Table 1, process If).

Figure 17 illustrates a less spectacular but quite common situation, namely degeneracies leading to super Coster-Kronig fluctuation and decay of hole levels, in the present

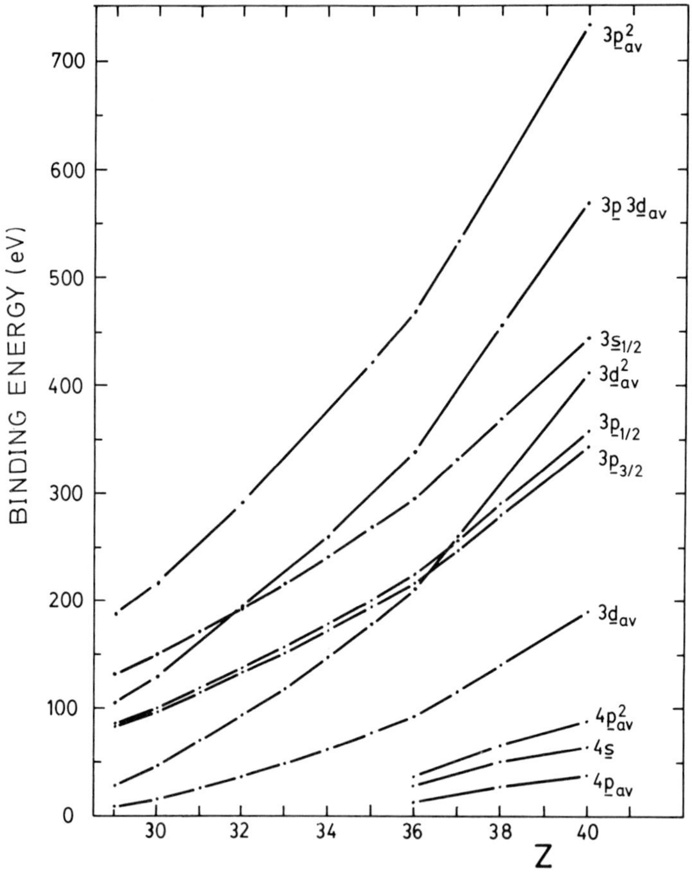

Fig. 17. Atomic, relativistic ΔSCF single-[82] and double-hole levels for $_{29}$Cu to $_{40}$Zr

case 3s and 3p levels in Kr and surrounding elements[66, 67] (Table 1, processes IIa, b). Since the final state electron does not belong to the same main shell as the three holes, the interaction strength is considerably weaker and the resulting shifts and widths are typically an order of magnitude smaller than in the giant Coster-Kronig case, i.e. 2–3 eV.

Finally, one should note that for heavy atoms the ΔSCF method quite generally fails to accurately describe core-level binding energies because of omnipresent *Coster-Kronig* fluctuation and decay processes (Table 1, entry III). Among the filled subshells of a given main shell, only the one with the highest angular momentum is basically unaffected and appears much narrower and much more unshifted than the corresponding deeper subshell hole levels. For instance, in the case of Hg[68], in the n = 4 region the 4s, 4p and 4d levels have widths and shifts (relative to ΔSCF) around 5–10 eV while the 4f-levels are essentially sharp and unshifted. Furthermore, the n = 5 region behaves similarly. Clearly, Coster-Kronig processes lead to rather small shifts in comparison with the binding energy, approximately equal to the line widths. Therefore, although in absolute numbers, the shifts and widths are large, this only represents a small modification of the one-electron picture in the ΔSCF sense.

5 Core-Hole Self-Energy $\Sigma_{\underline{i}}$ (E): Formal Theory and Approximative Treatment

So far, we have only discussed the general principles for core-hole relaxation and correlation and given a schematic picture of the resulting core-level spectra. However, it is of great interest to put the theory on a more quantitative form in order to interpret core-level spectra of the individual elements. Previously, many-body theory has been quite successful in interpreting the dynamics of atomic systems as observed in photoabsorption experiments ([14, 24, 25, 69)]and references therein), and one can expect this technique to be equally successful in the description of photoelectron spectra.

Since the dynamics of the core hole i is given by its self-energy $\Sigma_{\underline{i}}(E)$, the problem is how to calculate this quantity in a sufficiently accurate way. Clearly, in a region of level crossing, as discussed in Sect. 4, an accurate calculation is necessary even for a rough interpretation of an experimental spectrum. Furthermore, if the coupling strength is large, the range of effective level crossing can be very wide and include a large number of elements, probably $48 \leqslant Z \leqslant 58$–70. An accurate description of the core-hole self-energy must then go far beyond the second-order treatment in Sect. 3.1 (Eq. (12)).

The basic aspects of the improved treatment can be formulated in very simple terms. A reasonably reliable calculation of $\Sigma_{\underline{i}}(E)$ will have to take into account the following physical effects:

a) In the intermediate state (Fig. 9e), the core hole $\underline{j}$ will interact with and modify the electron-hole excitation $\underline{k}$m, which thus takes place in a singly ionized system. As a consequence, the excited electron m will feel the potential of two core holes (V^{N-2} potential). Furthermore, the two core holes $\underline{j}$ and $\underline{k}$ must be allowed to relax properly. This involves in principle a two-core hole self-energy which describes relaxation, correlation, decay, screening of the hole-hole repulsion etc. and it also affects the interaction with the excited electron m.

b) As discussed in Sect. 2.1 (Fig. 3) an electron-hole excitation may become strongly modified by the dielectric response of the system. If the response of the k-th shell is strongly collective for $\underline{k}$m type of excitations, this excitation channel has to be treated within the RPAE[14, 24, 25)] and, as a result, the excitation matrix element becomes *screened.* This can have a very pronounced effect on the distribution of spectral strength.

5.1 General Treatment of the Self-Energy $\Sigma_{\underline{i}}$ (E)

In this section I should like to outline a general approach to the *renormalization* (infinite order treatment) of the self-energy $\Sigma_{\underline{i}}(E)$ and describe how this scheme has been partly implemented in the actual calculations to be presented in Sect. 6. The general, formal theory of many-body systems and renormalization is well known (see e.g.[27, 70)] and references therein). However, in actual applications each physical system has its own characteristics and special problems and only recently has the full power of many-body theory begun to be applied to excitation spectra in atoms and molecules[14, 19, 24, 25, 30–32, 69, 71–79]. For readers interested in a detailed discussion of the general framework, there is an extensive review of the application of diagrammatic many-body theory and Green's function methods to molecules by Cederbaum and Domcke[30]. The following discussion of perturbation expansions for the core-hole self-energy $\Sigma_{\underline{i}}(E)$ will be based entirely on the second-order direct diagram (Fig. 9e), and with few exceptions, exchange processes will only be considered in actual calculations. Furthermore, we continue to illustrate the general principles using only diagrams that propagate entirely forwards in time, leaving all Fermi sea correlation effects to the detailed applications.

The essence of the renormalization treatment is to consider self-energy diagrams to an order just high enough to identify the characteristic interactions involved in the problem to be studied and then to replace the interactions by *effective interactions* representing an infinite order treatment. The expansion scheme is illustrated in Fig. 18. The core hole is dressed up by the self-energy through the Dyson equation (Fig. 18a; see also Fig. 9b). The second-order self-energy in Fig. 18b describes the $\underline{jk}$m ionic excitation within the ground-state HF picture, i.e. the particles and holes are unrelaxed, the excited electron m is a virtual HF orbital in a neutral atom (V^N) potential and the particles and holes do not interact with each other.

The simplest corrections to this picture enter in *third order* (Figs. 18c–e), describing repulsion between the holes (Fig. 18c) and attraction between the electron and the holes

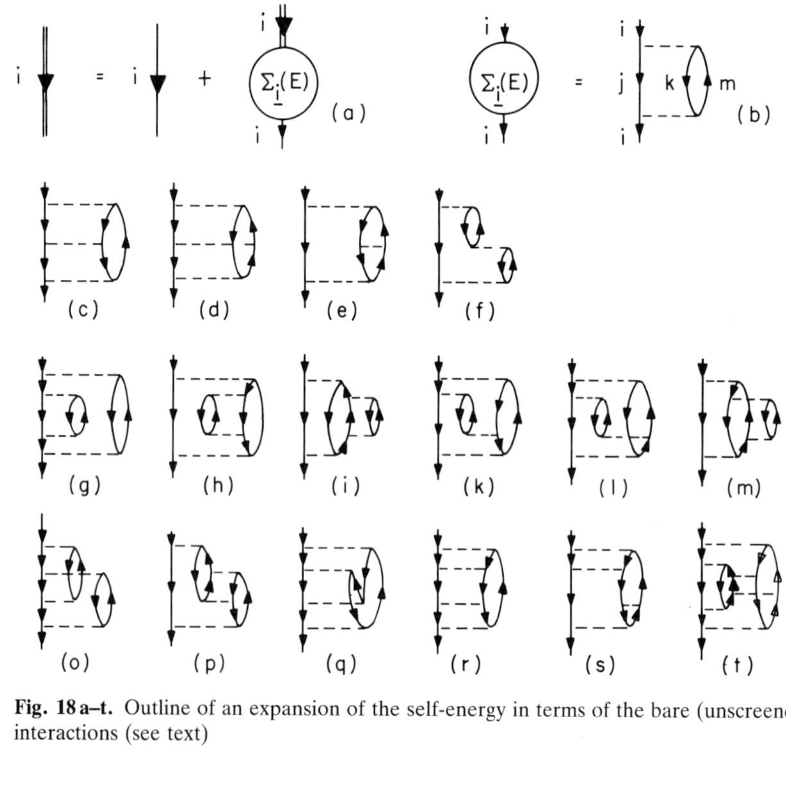

Fig. 18a–t. Outline of an expansion of the self-energy in terms of the bare (unscreened) Coulomb interactions (see text)

(Figs. 18 d, e). Among the third-order exchange processes, the most interesting diagram is Fig. 18 f, representing *density correlations* in the electronic medium in response the $\underline{i} \rightarrow \underline{j}$ core-hole transition. This diagram is the first in a sequence that constitutes the *random-phase approximation (RPA)*[21–25, 27–32]. Very often, Fig. 18 f is considered as the direct process and Fig. 18 e as its exchange and together they contribute to the *random-phase approximation with exchange* (RPAE)[14, 24, 25].

In *fourth order* (some typical diagrams are shown in Figs. 18 g–s) the first effects of *relaxation* of *the ionic excited states* enter. The self-energy part in Fig. 18 b now appears inside the self-energy itself and we see the first signs of the hierarchy of diagrams closing upon itself. Figs. 18 g–i then describe relaxations of the holes $\underline{j}$ and $\underline{k}$ and the electron m while Figs. 18 k–m (and many more) describe *screening* of the Coulomb interactions and Figs. 18 o–q complicated dynamic interactions. In addition, the fourth order contains straightforward iterations and mixtures of the third-order diagrams (e.g. Figs. 18 r, s).

It is now possible to see the general pattern. Proceeding from second order to infinite order there are classes of diagrams which only modify the individual electron and hole lines inside the second-order self-energy (Figs. 18 g–i and higher order iterations). The internal lines then become dressed themselves (Fig. 19 a with i = j, k) and we obtain the first step in the renormalization of the self-energy as shown in Fig. 19 b. However, as shown in third order (Figs. 18 c–f), the dressed electron and the dressed holes interact and all these effects can formally be included *exactly* by introducing an effective three-body interaction, as shown in Fig. 19 c. Therefore, Figs. 19 b, c describes the full self-enery in the most compact manner possible. Although this may look like hiding the difficulties, it is really a way of organizing the difficulties, collecting them into physically relevant quantities.

We shall now break down the renormalized self-energy diagram in Fig. 19 c in terms of effective two-body interactions suitable for explicit calculations. In analogy with the Dyson equation in Fig. 19 a, the total three-body, i.e. one-particle-two-hole (phh),

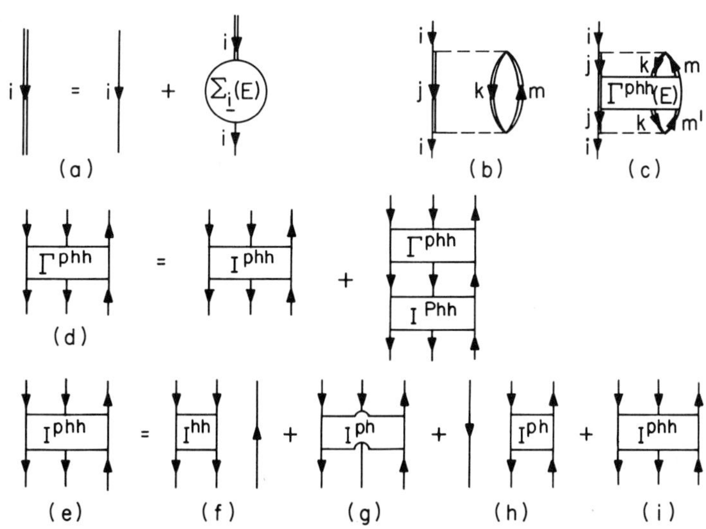

Fig. 19 a–i. Expansion of the self-energy in terms of effective interactions (see text)

interaction $\Gamma^{phh}(E)$ in Fig. 19c can be written in terms of an irreducible three-body interaction $I^{phh}(E)$ through a Bethe-Salpeter type of integral equation, as shown diagrammatically in Fig. 19d. $I^{phh}(E)$ describes the effective interaction between different one-particle-two-hole configurations and leads to energy level shifts and to a redistribution of the interaction strength. It also represents one of the main points where approximations are introduced. $I^{phh}(E)$ can be broken down into effective, energy-dependent two-body contributions (Figs. 19f–h) and true three-body blocks (Fig. 19i) where all three bodies are connected in a way more complicated than repeated two-body scattering (see e.g. Fig. 19t). There is an infinite hierarchy of equations coupling the motion of one body to the motion of two bodies, the motion of two bodies to three bodies etc.[23, 27, 30–32, 70)]. The problem is to truncate this chain in a physically meaningful way and in terms of quantities that can be calculated.

A simple and accurate way of truncation is to approximate the effective three-body interaction $I^{phh}(E)$ (Fig. 19e) by the sum of the effective two-body interactions $I^{hh}(E)$ and $I^{ph}(E)$ (Figs. 19f–h). The true three-body part $I^{phh}(E)$ only represents a small higher order correction (see e.g. Fig. 18t) and may be neglected. The second-order relaxation self-energy (Fig. 18b),

$$\Sigma_{\underline{i}}^{R}(E) = \mathop{\mathrm{S}}_{mjk} \frac{V_{jkim}\, V_{imjk}}{E_m^0 - E_{\underline{j}}^0 - E_{\underline{k}}^0 + E - i\delta} \tag{30a}$$

$$V_{jkim} = \langle jk|1/r_{12}|im\rangle \tag{30b}$$

can then be renormalized by introducing self-energy corrections to the intermediate holes $\underline{j}$ and $\underline{k}$ and including the effective (screened) hole-hole repulsion $I^{hh}(E)$ (Fig. 19f). The result can be written as a correction to the energy denominator in Eq. (30), as already briefly discussed in Sect. 3.4,

$$E_{\underline{j}}^0 + E_{\underline{k}}^0 \rightarrow E_{\underline{jk}}(E) = E_{\underline{j}}(E) + E_{\underline{k}}(E) - I_{\underline{jk}}^{hh}(E) \tag{31}$$

and includes the diagrams in Figs. 20b, c. The two-hole energy $E_{\underline{jk}}(E)$ is a complex quantity and represents in principle all effects of static and dynamic relaxation, decay and Fermi sea correlations in the doubly ionized system.

The self-energy can be renormalized further by inclusion of the diagrams in Figs. 20d–f which can be written in terms of an effective excitation matrix element as shown in Fig. 20g. Combining the renormalization of the energy denominator with the renormalization of the excitation strength, the self-energy is obtained in the form

$$\Sigma_{\underline{i}}^{R}(E) = \mathop{\mathrm{S}}_{mjk} \frac{V_{jkim}\, \Gamma_{imjk}(E)}{E_m^0 - E_{\underline{jk}}(E) + E - i\delta} \tag{32}$$

with the effective interaction Γ_{imjk} given by a Bethe-Salpeter integral equation (Fig. 20h)

$$\Gamma_{imjk}(E) = V_{imjk} + \mathop{\mathrm{S}}_{mjk} \frac{[I_{jmjm'}^{ph}(E) + I_{kmkm'}^{ph}(E)]\, \Gamma_{im'jk}(E)}{E_m^0 - E_{\underline{j'k}}(E) + E - i\delta} \tag{33}$$

Equations (32) and (33) are the basic formulae in our calculational procedure and represent essentially the same starting point as used by Cederbaum and coworkers[30)].

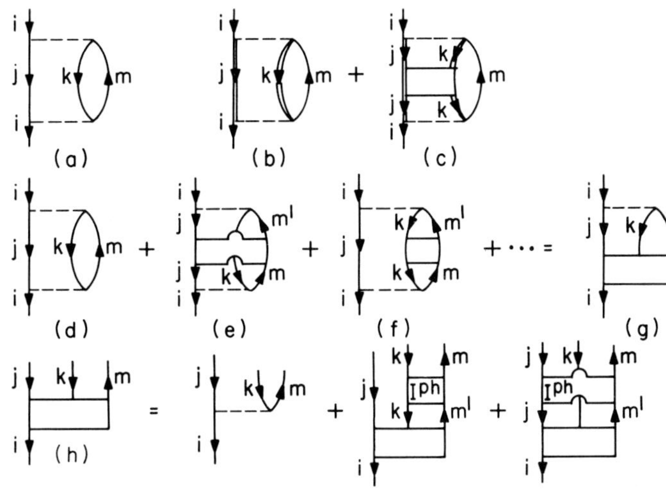

Fig. 20a–h. Expansion of the self-energy in terms of effective interactions (see text)

The differences are mainly concerned with the necessary further approximations of the effective two-body interactions $I^{ph}(E)$ and $I^{hh}(E)$ and the intermediate core-hole self-energies in $E_{j\underline{k}}(E)$. These approximations are intimately connected with the character of the zeroth-order spectrum of one-electron levels, whether the ionic excitation spectrum is mainly discrete or continuous, whether one is dealing with an atom or a molecule. Furthermore, as one goes to larger systems like large molecules or solids with small or vanishing valence-electron excitation energies, the diagram expansion has to be selected according to principles different from those used for finite systems[27)]. For instance, there is no longer any reason for discarding Fig. 18o while including Fig. 18g. Moreover, the time-reversed Fermi sea-correlation diagrams tend to become as important as the forward propagating ones discussed so far. We will illustrate some of these points in later sections.

5.2 Approximative Treatment of $\Sigma_{\underline{i}}^{R}(E)$

The further approximative treatment of the core hole relaxation self-energy can be formulated in the following way:

a) As discussed in Sect. 3.2, $\Sigma_{\underline{i}}^{R}(E)$ is divided into two parts corresponding to static relaxation (SR) and dynamic relaxation (DR)

$$\Sigma_{\underline{i}}^{R}(E) = \Sigma_{\underline{i}}^{SR}(E) + \Sigma_{\underline{i}}^{DR}(E) \tag{34}$$

b) The static part is approximated by monopole relaxation

$$\Sigma_{\underline{i}}^{SR}(E) \cong \Sigma_{\underline{i}}^{S0}(E) \tag{35}$$

and since we are not going to consider any monopole shake-up or shake-off structure, we approximate it further by the monopole relaxation shift of the single core hole and evaluate it by the HF or DF ΔSCF method

$$\Sigma_{\underline{i}}^{S0}(E) \simeq \Delta_{\underline{i}}^{S0} \quad \text{(a)}$$
$$\Delta_{\underline{i}}^{S0} \simeq E_{\underline{i}}^{HF}(\Delta SCF) - E_{\underline{i}}^{HF} \quad \text{(b)} \quad (36)$$
$$E_{\underline{i}}^{HF}(\Delta SCF) = E_{tot}^{HF}(\text{ground state}) - E_{tot}^{HF}(\text{hole in i}) \quad \text{(c)}$$

(c) Relativistic effects are introduced by replacing Hartree-Fock (HF) orbital energies by Dirac-Fock (DF) ones, taken from Desclaux[80])

$$E_{\underline{i}}^{HF} \rightarrow E_{\underline{i}}^{DF} \equiv E_{\underline{i}}^{0} \quad \text{(a)}$$
$$E_{\underline{i}}^{DF}(\Delta SCF) \simeq E_{\underline{i}}^{DF} + \Delta_{\underline{i}}^{S0} \quad \text{(b)} \quad (37)$$

Apart from these relativistic energy corrections, the treatment is non-relativistic regarding the theoretical formulation and one-electron wave function basis.

d) The above treatment is also applied to the case of two core holes in the same main shell. The self-energies and hole-hole repulsion in Eq. (31) are then approximated by static monopole relaxation and screening, evaluated through the ΔSCF method (cf. Sect. 3.6)

$$E_{\underline{i}}(E) \simeq E_{\underline{i}}^{0} + \Delta_{\underline{i}}^{S0} \quad \text{(a)}$$
$$E_{\underline{k}}(E) \simeq E_{\underline{k}}^{0} + \Delta_{\underline{k}}^{S0} \quad \text{(b)} \quad (38)$$
$$I_{\underline{jk}}^{hh}(E) \simeq F^{0}(j;k) - \Delta_{\underline{jk}}^{S0} \quad \text{(c)}$$

The Slater integral $F^0(j;k)$ describes the bare, average Coulomb interaction and $\Delta_{\underline{jk}}^{S0}$ the correction due to monopole screening. In the case of two core holes in the same subshell, we further have

$$\Delta_{\underline{jj}}^{S0} \simeq 2\Delta_{\underline{j}}^{S0} \quad (39)$$

This is because for j = k the diagrams in Figs. 18c, k give identical contributions in the monopole approximation. The presence of an extra core hole will only give small higher order corrections to the relaxation shift. For j, k belonging to the same main shell, it is still a reasonable approximation to set

$$\Delta_{\underline{jk}}^{S0} \simeq \Delta_{\underline{j}}^{S0} + \Delta_{\underline{k}}^{S0} \quad (40)$$

To check these results we have also performed explicit ΔSCF calculations for two core holes in a few cases

$$E_{\underline{jk}}(\Delta SCF) = E_{tot}^{HF}(\text{ground state}) - E_{tot}^{HF}(\text{holes in j, k}) \quad (41)$$

From the numerical results to be presented in Sect. 6 we shall find that non-linear effects are quite small and that the relaxation of two core holes can be rather accurately calculated from the relaxation of a single core hole.

e) The dynamic relaxation self-energy $\Sigma_{\underline{i}}^{DR}(E)$ is evaluated using an extended version of the renormalized second-order expressions Eq. (32) and (33)

$$\Sigma_{\underline{i}}^{DR}(E) = \mathrm{S} \frac{U_{jkim}\,\Gamma_{imjk}(E)}{E_m^0 - E_{\underline{jk}}(E) + E - i\delta} \tag{42}$$

$$\Gamma_{imjk}(E) = U_{imjk} + \mathrm{S} \frac{[I^{ph}_{jmjm'}(E) + I^{ph}_{kmkm'}(E)]\,\Gamma_{im'jk}(E)}{E_m^0 - E_{\underline{jk}}(E) + E - i\delta} \tag{43}$$

where U_{jkim} is defined to take into account important effects of time-reversed diagrams describing Fermi-sea correlation effects, which have not been discussed so far. The effective two-body interactions U, $I^{ph}(E)$ and $I^{hh}(E)$ (implicit in $E_{\underline{jk}}(E)$) in Eqs. (42) and (43) can be expanded in perturbation series. In Fig. 21 are shown those terms that have been included in actual calculations. The following points should be noted:

f) The screening of the hole-hole repulsion $I^{hh}_{\underline{jk}}(E)$ is calculated in the static monopole approximation and the double-hole energy $E_{\underline{jk}}(E)$ is obtained from Eq. (38), or equivalently Eq. (41). The double-hole level is therefore treated as sharp and structure-less.

g) Solving Eqs. (42) and (43) by omitting Figs. 21 h, k we obtain the response of the neutral free atom in the RPAE (random phase approximation with exchange; the present treatment is based on the formulation given in[24, 81]). However, the core hole $\underline{j}$ will perturb this response and by including the matrix elements in Figs. 21 h, k we obtain the RPAE response of the free atom perturbed by the average electrostatic potential of the core hole $\underline{j}$. Finally, by including the effect of the hole-hole repulsion and static monopole relaxation and screening (point (f)), the "atomic" response becomes shifted in energy.

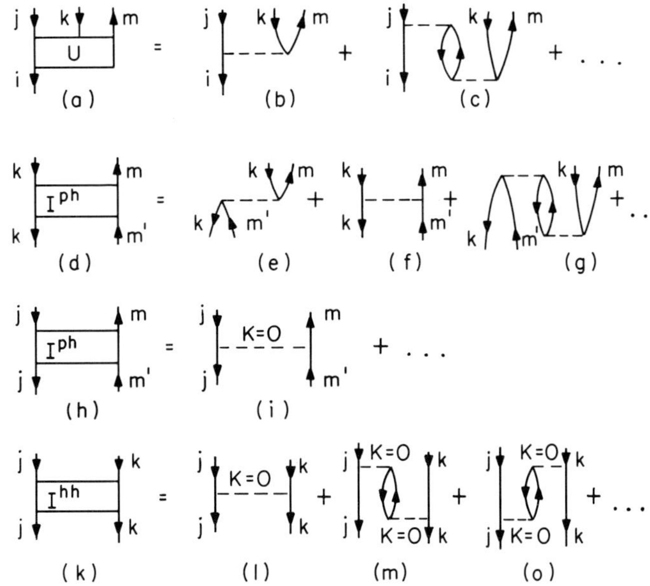

Fig. 21 a–o. Expansions of the effective interactions in terms of the unscreened Coulomb interactions (see text)

Still, this does not quite reproduce the true ionic response because screening of the electron-hole electrostatic interaction (Figs. 18 l, m) and relaxation of the excited electron have been neglected. Therefore, in all our calculations the excited electron m feels the V^{N-2} potential from the frozen HF-ground state with two electrons removed. This is a reasonable approximation if the main ionic excitation strength lies in the continuum and if the density of states varies slowly over distances comparable to the relaxation shift. However, the approximation becomes questionable when the variation of the continuum density of states is rapid (e.g. due to shape resonances) and it breaks down for low-lying, localized ionic excitations. In this case, one can explicitly estimate the effects of relaxation and screening within a single configuration. This is more difficult in the continuum case where one has to work with matrix elements of the electron self-energy and the electron-hole interaction between continuum states[24, 81)].

We have now formulated the approximation and presented one way of actually calculating it, namely by means of Eqs. (42), (43) and Fig. 21. Another way of including much of the higher order effects in Eq. (42) is to calculate the wave functions for the excited electron m in a potential which directly includes a certain selection of the interaction matrix elements in Fig. 21. This is going to be our normal mode of operation, and the zeroth-order basis set will be chosen in the following way:

a) Occupied (hole) states: Free atom HF ground-state orbitals.

b) Excited (particle) states: Two-core-hole (V^{N-2}) HF potential based on frozen ground-state orbitals. The excited electron is coupled to the the core hole $\underline{k}$ to a ^1P total angular momentum state (exactly like in photoionization[14, 24, 25, 69)]. The core hole $\underline{j}$ is a spectator, contributing only to the spherically symmetric Hartree potential

$$(\underline{j})_{av}(\underline{k}m\ ^1P)\ \mathrm{HF}\ V^{N-2} \tag{44}$$

Using this basis is the same as working in the TDAE (Tamm-Dancoff approximation with exchange[14, 24, 30, 71)]) for the $\underline{k}$m dipole response and including the Hartree potential from the second hole $\underline{j}$. This is rather close to what Cederbaum and coworkers call the 2ph-TDA[30)]. In order to arrive at the RPAE response, one has to include in particular the Fermi sea correlation diagrams in Figs. 21 c, g through the integral equation, Eq. (43), which can now be expanded to rather low order.

Finally, a comment regarding relativistic effects and the calculation of one-electron energies and monopole relaxation shifts. The most convenient way to obtain relativistic ΔSCF one-electron energies is to use the Dirac-Fock-Slater (DFS) ΔSCF values tabulated by Huang et al.[82)]. These are very close (a few tenths of an eV) to DF ΔSCF, and the relativistic monopole relaxation shift is the given by

$$\Delta_{\underline{i}}^{S0} = E_{\underline{i}}^{DF}(\Delta SCF) - E_{\underline{i}}^{DF} \tag{45 a}$$

$$\cong E_{\underline{i}}^{DFS}(\Delta SCF) - E_{\underline{i}}^{DF} \tag{45 b}$$

Equation (45) is to be preferred to the non-relativistic Eq. (36 b) in the case of heavy atoms with large spin-orbit slittings. $\Delta_{\underline{i}}^{S0}$ now measures monopole relaxation and screening including relativistic effects and is extremely useful for the calculation of approximate binding energies of double and multiple vacancy levels (Eq. (38)–(40)) for which ΔSCF results are very scarce.

6 Application to X-Ray Photoelectron Spectra in Free Atoms

So far, we have fairly extensively discussed the general aspects of static and dynamic relaxation of core holes. We have also discussed in detail methods for calculating the self-energy $\Sigma_{\underline{i}}(E)$. Knowing the self-energy, we know the spectral density of states function $A_{\underline{i}}(E)$ (Eq. (10)) which describes the X-ray photoelectron spectrum (XPS) in the sudden limit of very high photoelectron kinetic energy (Eq. (6)). We will now present numerical results for $\Sigma_{\underline{i}}(E)$ and $A_{\underline{i}}(E)$ and compare these with experimental XPS spectra and we will find many situations where atomic core holes behave in very unconventional ways.

6.1. The 4$\underline{s}$, 4$\underline{p}$ Core-Hole Spectrum in Xe and Surrounding Elements

The monopole relaxation shifts for single 4$\underline{s}$, 4$\underline{p}$ and 4$\underline{d}$ core holes calculated by the DFSΔSCF method (Eq. (45)) are shown in Fig. 22, and one should note the approximately linear Z-dependence which is to be expected. The corresponding DFSΔSCF energies[82] are shown in Fig. 15. The monopole relaxation shifts in Fig. 22 have also been used in Eqs. (38)–(40) to calculate relaxed two-core hole levels, some of which have also been calculated through the HFΔSCF method (Eq. (41)) including relativistic corrections. The resulting approximate DFΔSCF double-core hole energies are shown in Fig. 15.

In Fig. 23 we show the experimental results of Kowalczyk et al.[9, 10] and Gelius et al.[7, 8] for the elements $_{46}$Pd to $_{60}$Nd together with the results of two theoretical considerations: a) Most of the data are for *solids* and binding energies have to be corrected for work functions (i.e. referred to the vacuum level) and other solid-state effects like charge compression and changes in the relaxation shift due to the solid state environment. In order to introduce an atomic binding energy scale we note that the position of the 4$\underline{d}$ level in a free atom is well described by monopole relaxation and ground-state correlation

$$E_{4\underline{d}} \cong E^{0}_{4\underline{d}} + \Delta^{S0}_{4\underline{d}} + \Delta^{C}_{4\underline{d}} \tag{46}$$

The Fermi sea-correlation energy shift $\Delta^{C}_{4\underline{d}}$ is ~ -1 to -2 eV. From the experimental results for the 4$\underline{d}$ levels in[7–10] we then find the 4$\underline{d}$-level shift between the free atom and the solid

$$\Delta^{as}_{4\underline{d}} = E^{atom}_{4\underline{d}}(\text{theory; Eq. (46)}) - E^{solid}_{4\underline{d}}(\text{exp}) \tag{47}$$

$$\Delta^{as}_{4\underline{d}} = \Phi + \Delta^{s}_{4\underline{d}} \tag{48}$$

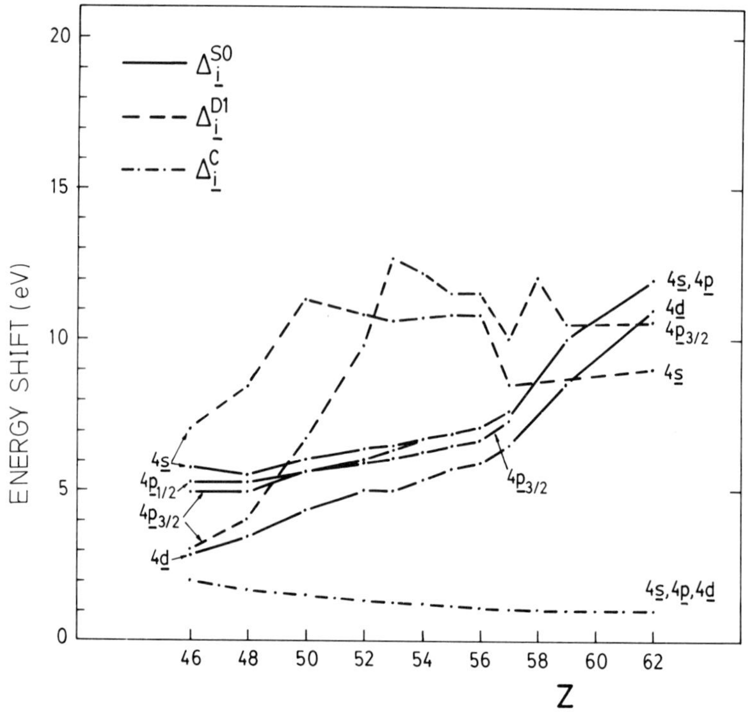

Fig. 22. Static and dynamic relaxation shifts in the range $_{46}$Pd to $_{62}$Sm. The static, monopole relaxation shift $\Delta_{\underline{i}}^{SO}$ (——) is a theoretical number (Eq. (45)). The Fermi sea correlation shift $\Delta_{\underline{i}}^{C}$ (–·–·–) is deduced by comparison with experimental 4$\underline{d}$ XPS spectra. The dynamic dipole relaxation shift $\Delta_{\underline{i}}^{D1}$ (–––) is deduced by comparison with experimental 4$\underline{p}$ XPS spectra (Eq. (49)), $\Delta_{\underline{i}}^{D1} = \Delta_{\underline{i}}^{DR}$)

Equation (48) represents a rough estimate of the shift due to solid-state effects $\Delta_{4\underline{d}}^{s}$ by subtracting the work function Φ from the total shift. Our results for the atom-solid shift $\Delta_{4\underline{d}}^{as}$ agree quite well with the theoretical and semi-empirical results of Johansson and Mårtensson[83].

The total atom-solid shift $\Delta_{4\underline{d}}^{as}$ is now assumed to be the same for all of the n = 4 single core levels and we introduce an *atomic scale* in the graphs in Fig. 23 by shifting the experimental scale by $\Delta_{4\underline{d}}^{as}$ to lower binding energy. Using this atomic scale we can now mark the free atom ΔSCF binding energies for the 4$\underline{s}$, 4$\underline{p}$, 4$\underline{d}^2$ and 4$\underline{p}$4$\underline{d}$ core levels and *deduce a residual shift which in principle must be due to non-monopole relaxation and fluctuation*. The experimental 4$\underline{s}$ and 4$\underline{p}$ core-hole energies read off the atomic scale should give a reasonable estimate of the free atom binding energies. As can be seen from Fig. 22, the difference

$$\Delta_{\underline{i}}^{DR} = E_{\underline{i}}^{atom}(\text{"exp"}) - E_{\underline{i}}(\Delta SCF) - \Delta_{\underline{i}}^{C} \tag{49}$$

is large, ranging form ~ 3 to 13 eV and *we identify this shift as being due almost entirely to the giant Coster-Kronig (gCK) fluctuation and decay processes*

$$4\underline{p} \leftrightarrows 4\underline{d}^2\,\mathrm{mf} \tag{50}$$

$$4\underline{s} \leftrightarrows 4\underline{p}\,4\underline{d}\,\mathrm{mf} \tag{51}$$

It is quite clear that when the gCK process reaches its maximum strength, the resulting shift $\Delta_{\underline{i}}^{DR}$ becomes a good deal larger than the normal static monopole relaxation shift $\Delta_{\underline{i}}^{S0}$.

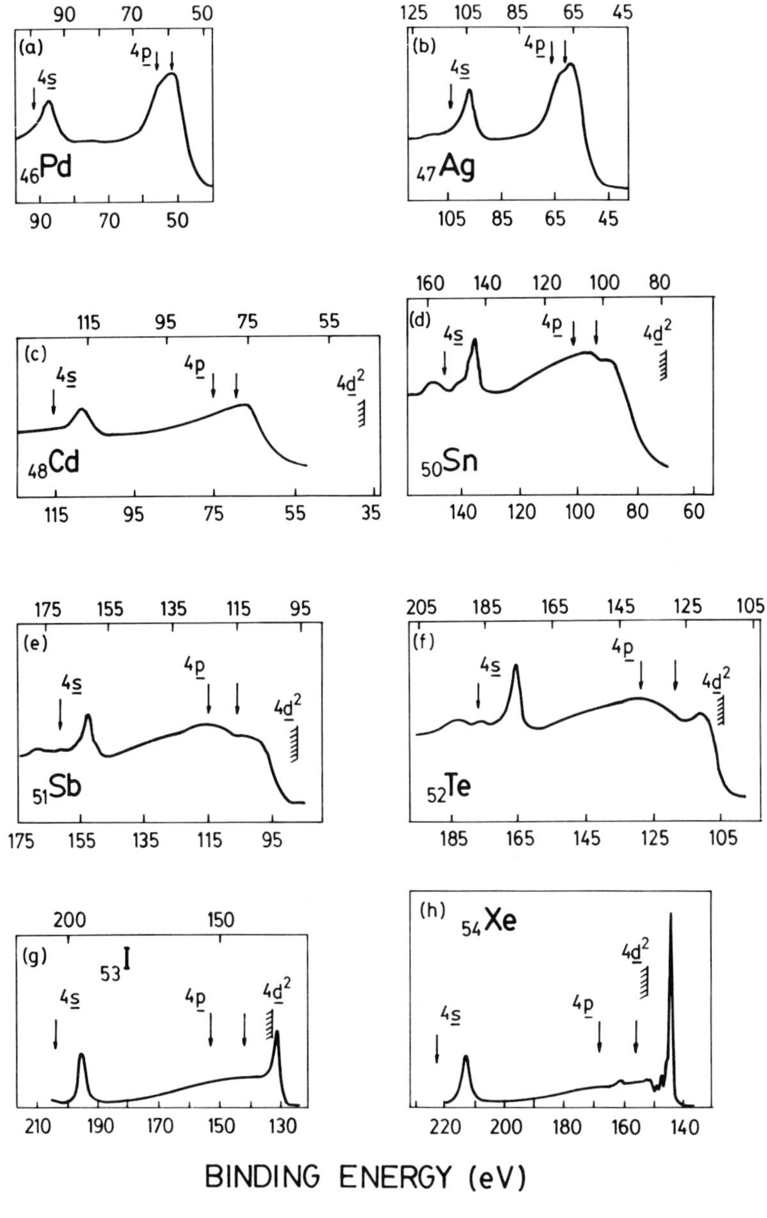

Fig. 23 a–h

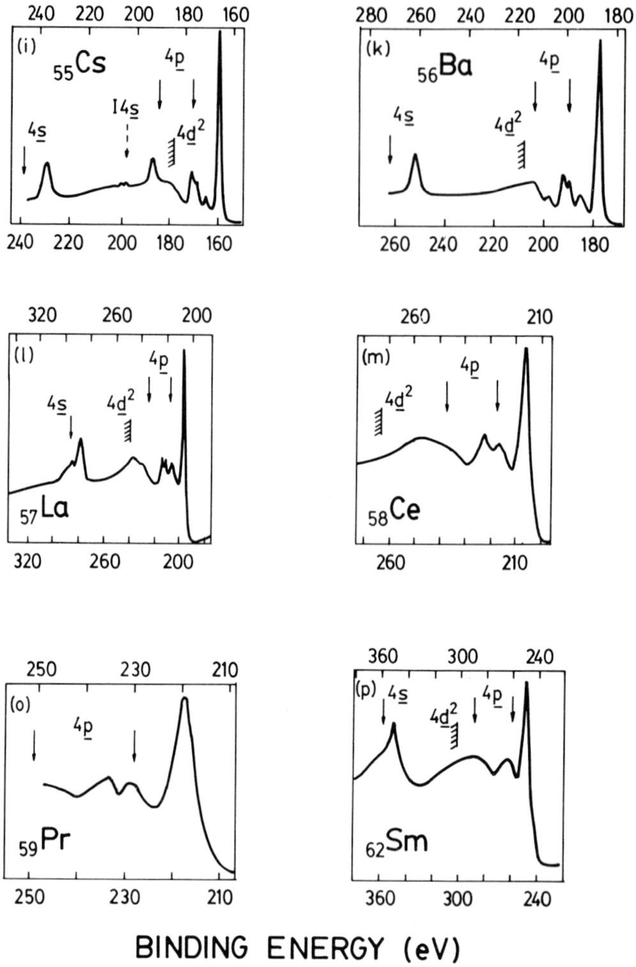

Fig. 23 i–p

Fig. 23 a–p. Experimental $4\underline{s}$, $4\,p$ X-ray photoelectron spectra: **(a)–(f)** and **(l)–(p)** by Kowalczyk and coworkers[9,10], **(g)–(k)** by Gelius and coworkers[7,8]. Atomic energy scales have been introduced (upper scales; see text) and atomic ΔSCF single-hole levels and double-hole thresholds have been marked. All spectra refer to metals except $_{53}$I (gaseous ICl), $_{54}$Xe (gas), $_{55}$Cs (CsI) and $_{56}$Ba ($Ba(NO_3)_2$)

Let us first discuss the $4\,\underline{p}$ case. The gCK fluctuation and decay process in Eq. (50) is illustrated in a one-electron level scheme in Fig. 24 a, and the corresponding second-order self-energy diagram for $\Sigma^{DR}_{4\underline{p}}(E)$ is shown in Figs. 24 b, c. Our technique for including higher order effects was discussed in detail in Sect. 5. Applied to the present $4\,\underline{p}$ case it can be briefly described in the follwing way:
The renormalized second-order self-energy takes the form

$$\Sigma^{DR}_{4\underline{p}}(E) \cong \mathop{S}_{m} \frac{U_m \Gamma_m(E)}{E^0_{mf} - E_{4\underline{d}^2} + E - i\delta} \tag{52}$$

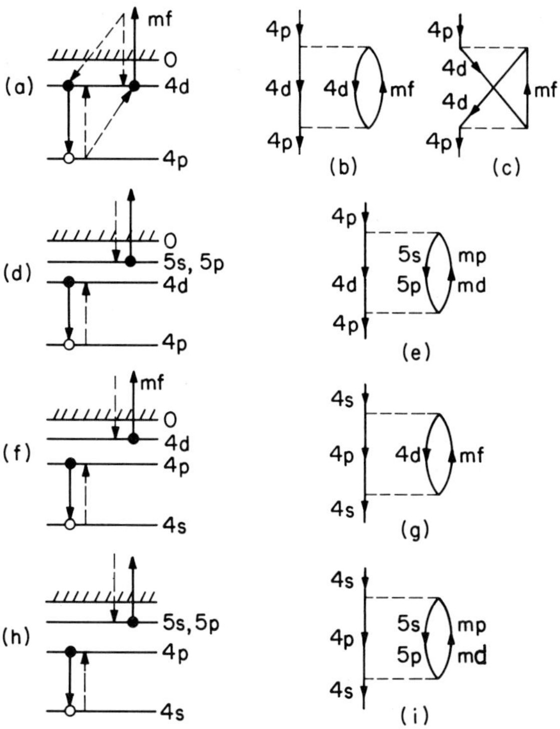

Fig. 24a–i. One-electron level pictures and self-energy diagrams for dynamic relaxation of 4s and 4p holes. **(a)–(c)** and **(f, g)** describe giant Coster-Kronig fluctuation (full plus dashed arrows) and decay processes (full arrows) while **(d, e)** and **(h, i)** correspondingly describe Coster-Kronig fluctuations and decay processes

In the second order, $U_m, \Gamma_m(E) \to V_m$ (Eq. (30b))

$$\sum_{m_\ell, m_s} U_m \Gamma_m(E) \to \sum_{m_\ell, m_s} |V_m|^2 = |g_m|^2 \tag{53a}$$

$$|g_m|^2 = \frac{14}{15} R^1(4d4d; 4p\,mf)^2 - \frac{8}{35} R^1(4d4d; 4p\,mf)\, R^3(4d4d; 4p\,mf) + \\ + \frac{32}{245} R^3(4d4d; 4p\,mf)^2 \tag{53b}$$

and to a good approximation we need only keep the dipole term

$$|g_m^1|^2 = \frac{14}{15} R^1(4d4d; 4p\,mf)^2 \tag{54}$$

Higher order corrections are then included according to Eqs. (42) and (43). The TDAE part of the atomic dielectric response including the effect of the Hartree potential of a spectator 4d hole is taken into account by our choice of basis set:

a) Occupied (hole) states: Free atom HF ground-state orbitals.

b) Excited (particle) states: V^{N-2} HF potential based on frozen atomic ground-state orbitals

$$(4d)_{av}(4d\,mf\ ^1P)\ \text{HF}\ V^{N-2} \tag{55}$$

The centre-of-gravity for the $4\underline{d}^2$ two-core-hole level is calculated with the HFΔSCF method plus relativistic corrections according to Eqs. (38)–(40), giving

$$E_{4\underline{d}^2} \cong 2\,E_{4\underline{d}}(\Delta SCF) - (F^0(4d;4d) - 2\Delta_{4\underline{d}}^{S0}) \tag{56a}$$

$$\cong 2\,E_{4\underline{d}}^{DF} - F^0(4d;4d) + 4\Delta_{4\underline{d}}^{S0} \tag{56b}$$

For some elements we also used the ΔSCF method involving two core holes (Eq. (41)). The results are shown in Fig. 15.

We now have the tools for constructing the *spectral function* $A_{4p}(E)$. We begin by studying the energy dependence of the self-energy and the solution to the Dyson equation, keeping the static monopole and the dynamic dipole terms and neglecting Fermi sea correlations

$$E - E_{4\underline{p}_j}^0 = \mathrm{Re}\,\Sigma_{4\underline{p}}^R(E) + \mathrm{Re}\,\Sigma_{4\underline{p}}^C(E) \tag{57a}$$

$$\cong \Delta_{4\underline{p}}^{S0} + \mathrm{Re}\,\Sigma_{4\underline{p}}^{D1}(E) \tag{57b}$$

$$E - E_{4\underline{p}_j}(\Delta SCF) \cong \mathrm{Re}\,\Sigma_{4\underline{p}}^{D1}(E) \tag{58}$$

$\mathrm{Re}\,\Sigma_{4\underline{p}}^{D1}(E)$, $\mathrm{Im}\,\Sigma_{4\underline{p}}^{D1}(E)$ and the straight lines $E - E_{4\underline{p}_j}(\Delta SCF)$ ($j = 1/2, 3/2$) are shown for a number of representative elements in Fig. 25, where the development towards breakdown of the quasi-particle picture for $4\underline{p}$ core holes can be clearly seen. Solutions corresponding to proper quasi-holes can only appear for crossings at the negative slope wings of $\mathrm{Re}\,\Sigma_{4\underline{p}}^{D1}(E)$, providing that the corresponding value of $\mathrm{Im}\,\Sigma_{4\underline{p}}^{D1}(E)$ is not too large

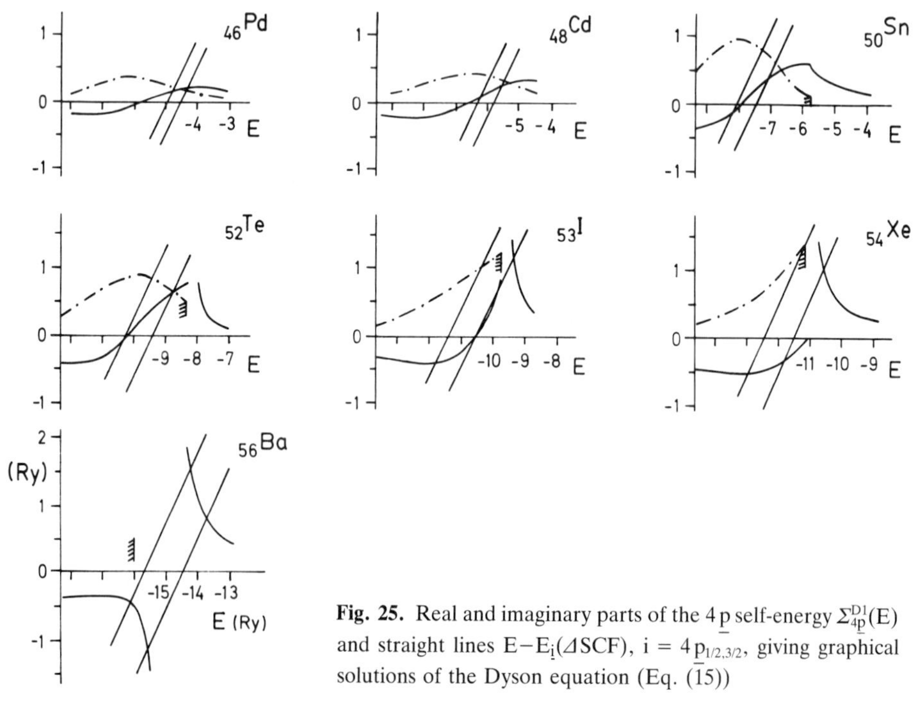

Fig. 25. Real and imaginary parts of the $4\,\underline{p}$ self-energy $\Sigma_{4\underline{p}}^{D1}(E)$ and straight lines $E - E_{\underline{i}}(\Delta SCF)$, $i = 4\,p_{1/2,3/2}$, giving graphical solutions of the Dyson equation (Eq. (15))

(in principle, one should look for the quasi-particle poles out in the complex E-plane). Crossings in the intermediate region where Re $\Sigma_{4\underline{p}}^{D1}(E)$ has a positive slope need not correspond to physical solutions. They occur where Im $\Sigma_{4\underline{p}}^{D1}(E)$ is very large and also $Z_{4\underline{p}} > 1$ and do not necessarily correspond to simple poles, or any poles at all, in the complex E-plane. The quasi-particle parameters are given in Table 2. A detailed discussion of the character of the solutions will be given in Sect. 6.1.2.

Table 2. Quasi-particle parameters for $4\underline{s}$, $4\underline{p}$ hole levels and spectra in the atomic elements $_{46}$Pd to $_{56}$Ba

Atom	Level[a]	F/R[b]	Present theory E (eV)[c]	Γ (eV)[d]	Z[e]	Comment[f]	Experiment E (eV)[l]	Γ	(eV)
$_{46}$Pd	$4\underline{p}_{3/2}$, $4\underline{p}_{1/2}$	F	−58.2	10.0	–	non-Lorentzian	−58[g]	10[g]	6.5[h] 6.5[h]
	$4\underline{s}_{1/2}$	F	−92.3	4.5	0.89	~ Lorentzian	−94[g]	4.8[g]	5.0[h]
$_{48}$Cd	$4\underline{p}_{3/2}$, $4\underline{p}_{1/2}$	F	−74.9	16.7	–	non-Lorentzian	−76[g]	13[g]	10[h] 12[h]
	$4\underline{s}_{1/2}$	F	−115.6	3.5	0.85	~ Lorentzian	−117[g]	4.1[g]	4.0[h]
$_{50}$Sn	$4\underline{p}_{3/2}$	F	−95.8	29	–	breakdown of	−96[g]	22[i]	17[h]
	$4\underline{p}_{1/2}$	R	−97.9	15.0	–	Q. P. P.			17[h]
	$4\underline{s}_{1/2}$	F	−143.8	0.7	0.43	~ Lorentzian	−144[g]	3.2[g]	2.8[h]
$_{52}$Te	$4\underline{p}_{3/2}$	F	−116.8	34	–	breakdown of	−121[g]	26[i]	
	$4\underline{p}_{1/2}$	R	−120.7	16.0	–	Q. P. P.			
	$4\underline{s}_{1/2}$	F	−176.5	1.8	0.7	~ Lorentzian	−176[g]	3.0[g]	
$_{53}$I	$4\underline{p}_{3/2}$	F R	−131.1 −133.9			Q. P. P. possibly valid	−129.5[k]	(2.8)[k]	
	$4\underline{p}_{1/2}$					breakdown of Q. P. P.			
	$4\underline{s}_{1/2}$	F	−196.5	2.8	0.7	~ Lorentzian	−195.4[k]	2.7[k]	
$_{54}$Xe	$4\underline{p}_{3/2}$	F R	−145.5 −148.5	0.8 1.0	0.42 0.54	non-Lorentzian	−145.1[k]	0.60[k]	
	$4\underline{p}_{1/2}$					breakdown of Q. P. P.			
	$4\underline{s}_{1/2}$	F	−214.7	2.8	0.8	~ Lorentzian	−213.3[k]	2.9[k]	
$_{56}$Ba	$4\underline{p}_{3/2}$	F	−184	1.3	0.65	~ Lorentzian	−187[k]	2.2[k]	
	$4\underline{p}_{1/2}$	F	−220	5	0.7	non-Lorentzian	broad structure (exp. on solid)		

a) Denotes character of unperturbed (ΔSCF) levels, not orbital character of actual spectral peaks
b) F = frozen, R = relaxed $4\underline{d}^2/4\underline{p}\,4\underline{d}$ ionic core (see text)
c) Position of maximum in *total* spectral function (Figs. 26, 30)
d) Full width at half maximum (FWHM) of *total* spectral function (Figs. 26, 30). Superposition of $4\underline{p}_{1/2}$ and $4\underline{p}_{3/2}$ spectra
e) From Eq. (17). Dash (−) denotes that the simple analysis in Eqs. (15)–(19) does not apply
f) Q. P. P. = Quasi-particle picture
g) Ref.[9, 10]; h) Ref.[84]; i) Ref.[213]; k) Ref.[7, 8]
l) Read off the *atomic* scale, i.e. corrected for the atom-solid shift (Eq. (47))

Our knowledge of the self-energies in Fig. 25 now enables us to construct the corresponding spectral functions $A_{4\underline{p}_{1/2}}(E)$ and $A_{4\underline{p}_{3/2}}(E)$ according to Eq. (10), and we also define a statistically weighted total spectral function

$$A_{4\underline{p}}(E) = \frac{1}{3} A_{4\underline{p}_{1/2}}(E) + \frac{2}{3} A_{4\underline{p}_{3/2}}(E) \tag{59}$$

which is closely related to the X-ray photoelectron spectrum (Eq. (6)) and can be compared with experiment. The results for the total 4 p spectral function in Eq. (59) are shown in Fig. 26.

In the corresponding case of a 4 $\underline{s}$ core hole level there are a number of complications, namely the strong effects of spin-orbit splitting and dynamic relaxation of the 4 $\underline{p}$ hole in the 4 $\underline{p}$ 4 $\underline{d}$ configuration. However, if we are mainly interested in the fluctuation shift of

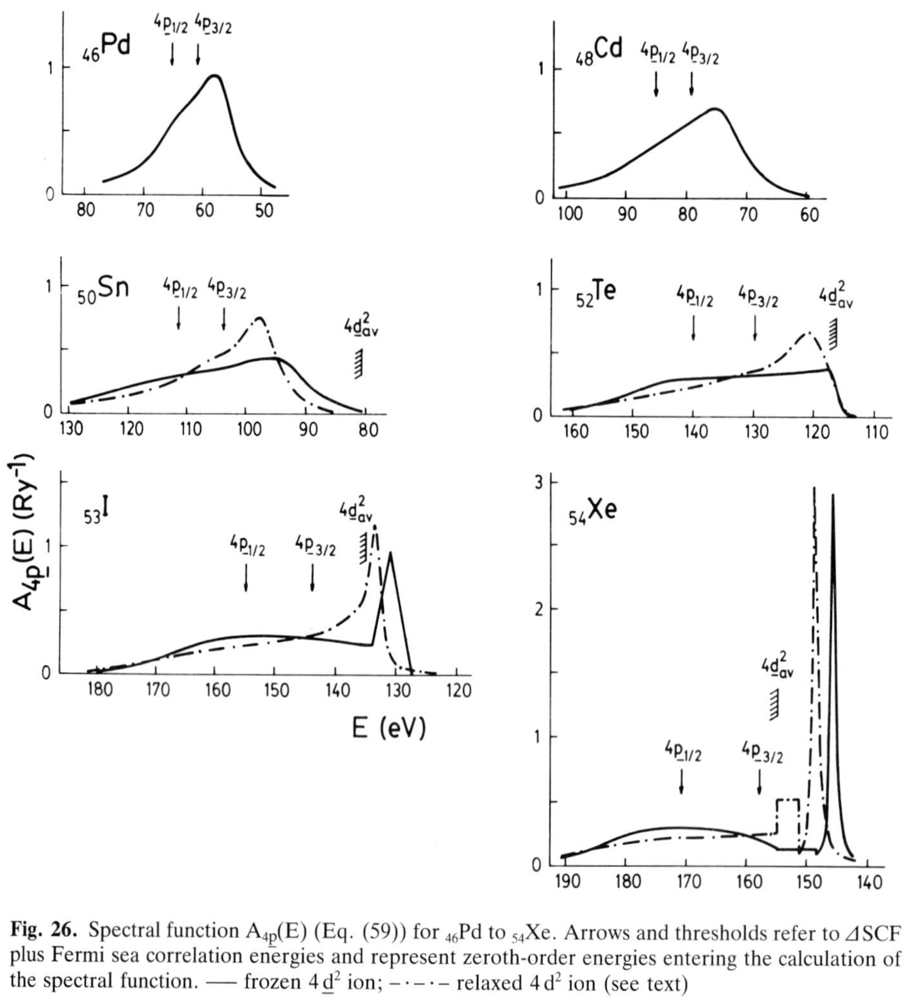

Fig. 26. Spectral function $A_{4\underline{p}}(E)$ (Eq. (59)) for $_{46}$Pd to $_{54}$Xe. Arrows and thresholds refer to ΔSCF plus Fermi sea correlation energies and represent zeroth-order energies entering the calculation of the spectral function. —— frozen 4 $\underline{d}^2$ ion; –·–·– relaxed 4 d^2 ion (see text)

the 4s quasi-hole, then the details of the threshold structure are integrated over and need not be considered in a first approximation. Thus, in order to really simplify this problem we use DFΔSCF for the threshold energy[19]. In analogy with the 4p case, the 4s self-energy takes the form

$$\Sigma_{4\underline{s}}(E) = \Delta_{4\underline{s}}^{S0} + \Sigma_{4\underline{s}}^{D1}(E) \tag{60}$$

$$\Sigma_{4\underline{s}}^{D1}(E) = \mathop{\mathrm{S}}_{m} \frac{U_m^1 \Gamma_m^1(E)}{E_{mf}^0 - E_{4\underline{p}4\underline{d}} + E - i\delta} \tag{61}$$

where, in second order, the coupling strength distribution becomes

$$|g_m^1|^2 = 2\,R^1\,(4p4d;\ 4s\ mf)^2 \tag{62}$$

The inclusion of higher order corrections is performed as in the 4p-case. Graphical representations of the real and imaginary parts of the 4s self-energy and solutions of the Dyson equation are shown in Fig. 27, and the corresponding quasi-particle parameters are given in Table 2.

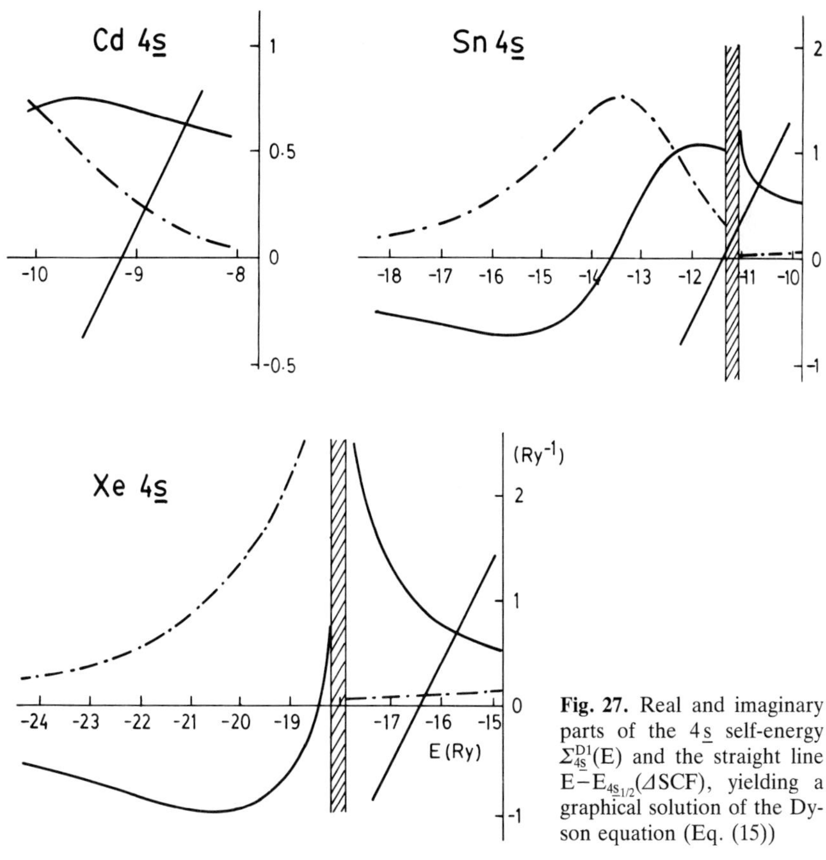

Fig. 27. Real and imaginary parts of the 4s self-energy $\Sigma_{4\underline{s}}^{D1}(E)$ and the straight line $E - E_{4\underline{s}_{1/2}}(\Delta SCF)$, yielding a graphical solution of the Dyson equation (Eq. (15))

As seen in Table 2, for those solutions which occur outside the gCK continuum the width is quite small, being determined by the much weaker Coster-Kronig and Auger processes. For a number of elements we have included the dipole parts of the following Coster-Kronig fluctuation and decay processes:

4p, Figs. 26 d, e: $4\underline{p} \leftrightarrows 4\underline{d}5\underline{p}md$ (63 a)

$$|g^1_{m}|^2 = \frac{8}{9} R^1(4d5p; 4p\,md)^2 \tag{63 b}$$

$4\underline{p} \leftrightarrows 4\underline{d}5smp$ (63 c)

$$|g^1_{m}|^2 = \frac{4}{9} R^1(4d5s; 4p\,mp)^2 \tag{63 d}$$

4s, Figs. 26 h, i: $4\underline{s} \leftrightarrows 4\underline{p}5\underline{p}md$ (64 a)

$$|g^1_{m}|^2 = \frac{4}{3} R^1(4p5p; 4s\,md)^2 \tag{64 b}$$

$4\underline{s} \leftrightarrows 4\underline{p}5\underline{s}mp$ (64 c)

$$|g^1_{m}|^2 = \frac{2}{3} R^1(4p5s; 4s\,mp)^2 \tag{64 d}$$

These processes were originally included in order to broaden the 4s and 4p levels so that the discrete lines could be treated in the same manner as the continuum, without any intention of accurately describing the line width. However, the approach gave quite satisfactory results with the following approximations[19]:

a) HF (Koopman's) energies for the double-vacancy levels, however including a 1 Ry fluctuation shift of any intermediate 4p hole.
b) Excited electron wave functions from photoabsorption calculations with a spherically averaged core hole, like e.g.

$$(5\underline{p}md)_{av} HF\ V^{N-1} \tag{65}$$

Since the energy of the electron in the Coster-Kronig decay is high up in the continuum (cf. Fig. 2) one can argue that the choice of basis set is not very critical and that neglecting the attraction of the extra hole is compensated for by neglecting the repulsion from coupling the electron-hole excitation to ^1P.

The 4s, 4p problem has recently been reinvestigated in greater detail for the entire range of elements Pd to Xe by Ohno and Wendin[85], confirming and extending the conclusions in[19]. The results in Figs. 26, 27, 30 and Table 2 are partly taken from[85]. It should be noted that near the crossing of the 4s and 4p 4d levels around $_{50}$Sn (Fig. 15), the approximation making use of a single, sharp, average 4p 4d threshold must break down, necessitating a description in terms of a continuum distribution of 4p 4d thresholds caused by the gCK fluctuation and decay (cf. the 4p core hole spectrum in Sn-Te in Fig. 26). As a result, the giant Coster-Kronig decay of a 4s hole in Sn will probably not be forbidden, which would lead to a considerably larger width than the Coster-Kronig width in Table 2.

Furthermore, effects of relaxation of the $4\underline{d}^2$ and $4\underline{p}\,4\underline{d}$ vacancies on the final state continuum f-electron have been considered. This can be formulated in terms of screening of the attractive potential felt by the f-electron due to the double vacancy (cf. the $4d \rightarrow \varepsilon f$ photoabsorption case[24, 81]), representing, however, a fairly involved procedure. In order to estimate the effects of relaxation, a reasonable way appears to calculate the final state f-electron in the potential of the relaxed $4d^9$ ($4p^5$) ion with an additional hole in the 4d-shell, giving a V^{N-2} potential[85, 86]. The f-electron then will feel the attractive potential of the double vacancy with the inclusion of relaxation due to one of the holes. Loosely speaking, this represents a transition type of state, and the results for the spectral function of a $4\underline{p}$ hole (Fig. 26) are in agreement with experiments. This is particularly clear in the case of $L\gamma_{2,3}$ ($2\underline{p} \rightarrow 4\underline{s}$) X-ray emission spectra for Pd to Xe[86], where there is no disturbing inelastic background. In the XPS case (Fig. 26), inelastic scattering of the escaping photoelectrons appears to play a very important role in the solids Sn to Te. This problem is further discussed in [85, 86] where a tentative comparison of theory with experiment is given.

6.1.1 *Disussion of the Validity of the Treatment in the Case of Ba*

As discussed in Sect. 5, our approximation scheme for the self-energy $\Sigma_{\underline{i}}(E)$ is designed for a situation where the main ionic excitation strength lies in the continuum. However, from Fig. 25[16] it is obvious that this situation is not at hand in Ba because the main ionic excitation strength (in this approximation) has become concentrated to the $4\underline{d}_{av}(4\underline{d}\,4f\,^1P)$ level. The approximation will then have to be improved in typically two ways:

(a) The *monopole relaxation* of the $4\underline{d}^2 4f$ configuration has to be calculated in a self-consistent manner. This means that the $4\underline{d}$-$4f$ Coulomb attraction has to be screened, in the same way as previously discussed for the $4\underline{d}$-$4\underline{d}$ repulsion, and also the 4f self-induced static monopole relaxation shift has to be included.

(b) The *multiplet structure* of the $4\underline{d}^2 4f$ configuration has to be treated in a more careful way.

In the case of a single $4\underline{d}^2 4f$ configuration one can apply the full renormalization treatment as discussed in Section 5.1 and shown in Figs. 28a, b. Including only the effective two-body interactions, the excitation energy of the $4\underline{d}^2 4f$ configuration can be written in general terms as

$$E_{4\underline{d}^2 4f}(E) = E_{4f}(E) - E_{4\underline{d}}(E) - E_{4\underline{d}}(E) + I^{hh}_{4\underline{d}4\underline{d}}(E) - I^{ph}_{4\underline{d}4f}(E) - I^{ph}_{4\underline{d}4f}(E) \tag{66}$$

The energies and interactions in Eq. (66) can also be regarded as effective operators including the effects of all other configurations. Including relaxation and screening in the static monopole approximation one obtains

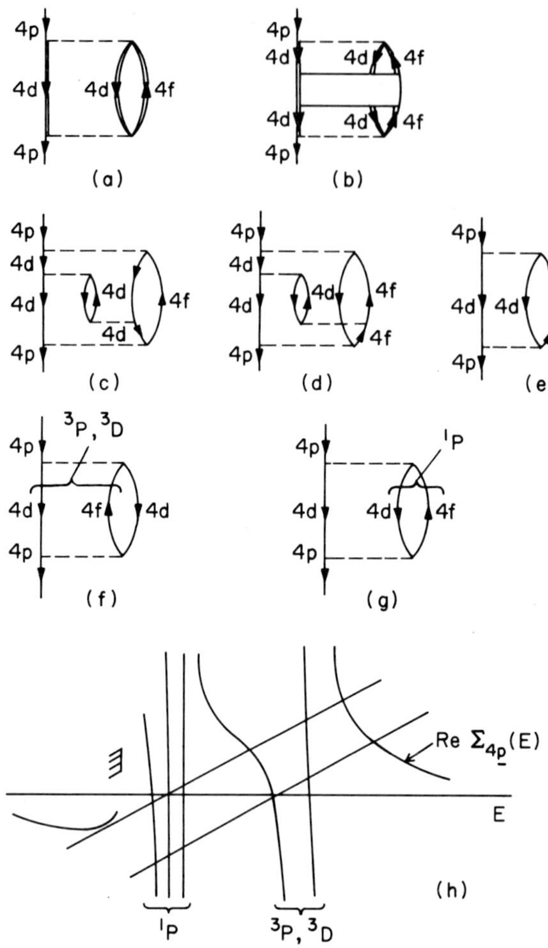

Fig. 28 a–h. 4 p self-energy diagrams in a case where a $4\underline{d}^2 4f$ ionic excited level dominates the excitation spectrum (see text and compare with Figs. 10, 17–19)

$$E_{4f}(E) \cong E_{4f}^{HF} - \Delta_{4f}^{S0} \quad \text{(here HF = virtual HF)} \tag{67 a}$$

$$E_{4\underline{d}}(E) \cong E_{4\underline{d}}^{0} + \Delta_{4\underline{d}}^{S0} \tag{67 b}$$

$$I_{4\underline{d}4\underline{d}}^{hh}(E) \cong F^0(4\,d; 4\,d) - 2\Delta_{4\underline{d}}^{S0} \tag{67 c}$$

$$I_{4\underline{d}4f}^{ph}(E) \cong F^0(4\,d; 4\,f) - \Delta_{4\underline{d}4f}^{S0} \tag{67 d}$$

Inserting Eqs. (67 a–d) into Eq. (66) gives the approximate ΔSCF result

$$E_{4\underline{d}^2 4f}(\Delta \text{SCF}) \cong E_{4f}^{0} - E_{4\underline{d}^2}(\Delta \text{SCF}) + \Delta E_{4f}^{S0} + \Delta E(\text{SL}) \tag{68}$$

where

$$E_{4f}^{0} = E_{4f}^{HF} - 2F^{0}(4d;4f) \tag{69a}$$

$$E_{4\underline{d}^2}(\Delta SCF) = 2E_{4\underline{d}}^{0} - F^{0}(4d;4d) + 4\Delta_{4\underline{d}}^{SO} \tag{69b}$$

$$\begin{aligned} \Delta E_{4f}^{SO} &= 2\Delta_{4\underline{d}4f}^{SO} - \Delta_{4\underline{d}}^{SO} \\ &\cong 2\Delta_{4\underline{d}}^{SO} + \Delta_{4f}^{SO} \end{aligned} \tag{69c}$$

and where $\Delta E(SL)$ represents the multiplet splitting of the $4\underline{d}^2 4f$ configuration. Equation (68) directly connects with the previous approach in Sect. 6.1 where the first, second and fourth terms have already been included or considered: The zeroth-order basis (Eq. (55)) is such that an excited electron feels a frozen HF V^{N-2}potential, which accounts for Eq. (69 a). Furthermore, the relaxed double-hole energy $E_{4\underline{d}^2}(\Delta SCF)$ is the same as before. Finally, for a single discrete configuration, the effective interaction $\Gamma(E)$ in Eqs. (42, 43) becomes an effective energy shift due to the residual interaction, and this goes into the multiplet splitting term $\Delta E(SL)$ in Eq. (68). The *new contribution* ΔE_{4f}^{SO} describes the effect of the inclusion of static monopole screening of the $4\underline{d}$-4f interaction (Figs. 28 c, d) and static monopole relaxation of the 4f electron (Fig. 28 e), and allows the 4f electron to influence the relaxation process of the entire configuration. As seen from Table 3, the 4f orbital is almost as compact as the 4d orbital and therefore the 4f static monopole relaxation shift Δ_{4f}^{SO} must be nearly as large as $\Delta_{4\underline{d}}^{SO}$, say $\Delta_{4f}^{SO} \cong 0.3$ Ry, leading to a very large relaxation correction $\Delta E_{4f}^{SO} \cong 1.2$ Ry.

The different results for the $4\underline{d}^2 4f$ energy levels are summarized in Table 3. The level given by the zeroth-order basis (Eq. (55)) corresponds to case (e), and if the level is made self-consistent it becomes pushed up by ΔE_{4f}^{SO} to the vicinity of the $4\underline{d}^2$ threshold. Obviously, a single configuration treatment then breaks down, and case (f) therefore represents a pseudo-state describing an average over the actual levels.

Case (d) in Table 3 shows the estimated ΔSCF energy for the $4\underline{d}^2 4f$ configuration average. However, the configuration average has only limited physical significance because of the very strong multiplet splitting. The span of the splitting is determined by

Table 3. Energy levels for the ionic excited $4\underline{d}^2 4f$ level in atomic Ba in various approximations illustrating the importance of self-consistency in the case of strongly localized levels

Case	Level	Comment	$-E_{4\underline{d}^2 4f}$ (Ry)		E_{4f} (Ry)	Radial overlap
(a)	$4\underline{p}_{3/2}$	ΔSCF		−14.5		
(b)	$4\underline{p}_{1/2}$	ΔSCF		−15.7		
(c)	$(4\underline{d}^2 4f)_{av}$	frozen	−15.5	−15.0	−2.7	0.89
(d)	$(4\underline{d}^2 4f)_{av}$	ΔSCF	−15.0	−14.5		
(e)	$(4\underline{d})_{av}(4\underline{d}\,4f\,^1P)$	frozen	−15.5	−15.0	−1.1	0.76
(f)	$(4\underline{d})_{av}(4\underline{d}\,4f\,^1P)$	ΔSCF	−16.7	−16.2	0	
(g)	$4\underline{d}_{av} + (4\underline{d}\,4f)_{av}$	[a]	−14.8	−14.3		
(h)	$(4\underline{d}^2 4f)_{av}$	[b]	−15.1	−14.6		

[a] $-E_{4\underline{d}^2 4f} = E_{4\underline{d}}(\Delta SCF) - \omega_{4\underline{d}4f}$, where $\omega_{4\underline{d}4f}$ is the $4d \rightarrow 4f$ excitation energy in the neutral atom ($V^{N-\overline{1}}$ HF potential)

[b] Relaxation of one single $4\underline{d}$ hole. The 4f electron is assumed to completely screen the other $4\underline{d}$ hole

the $G^1(4d; 4f)$ Slater exchange integral (cf. photoabsorption[14, 24]) and the relaxed $(4\underline{d}^2 4f)_{av}$ configuration should be split into two main groups of lines, one around the $(4\underline{d})_{av}(4\underline{d}\,4f\,^3P, ^3D)$ levels (Figs. 28f, h) and one around the $(4\underline{d})_{av}(4\underline{d}\,4f\,^1P)$ levels (Figs. 28g, h). The ^3P, ^3D parent levels should lie around the configuration average level while the levels based on the ^1P parent should become pushed up to the vicinity of the $4\underline{d}^2$ threshold (case (f)), as suggested in Fig. 28g. This is the collective shift of excitation strength as described by the TDAE[14, 24, 30]. However, ground-state correlations are very important and one therefore has to work in the RPAE[14, 24, 25, 30], as discussed in Section 5.2. This leads to a reduction of the singlet-triplet splitting, and one has then good reasons for expecting some prominent $(4\underline{d})_{av}(4\underline{d}\,4f\,^1P)$ levels below the $4\underline{d}^2$ threshold.

The reason why the previous treatment (Section 6.1, Fig. 25) gave reasonable results for the $4\underline{p}$ level shift can now be easily understood: The effective level (case (e)) lies with the correct weight approximately at the centre-of-gravity of the multiplet structure (Fig. 28g) and simulates the correct spectral repulsion.

As a final comment, the above discussion describes what is intuitively fairly clear, namely that a very compact electron-hole excitation is only weakly perturbed by the spherical part of an "external" potential. A reasonable approximation would then be to add the $4\underline{d}\,4f$ excitation energy in the neutral atom to the binding energy of a $4\underline{d}$ hole (Table 3, case (g)). Another way would be to start from the frozen $4\underline{d}^2 4f$ configuration average (case (c)) and argue that because the 4f electron is almost as compact as the $4\underline{d}$ hole, the net effect is that of a single hole and the total relaxation shift should be given by $\Delta^{SO}_{4\underline{d}}$. The result is shown in Table 3, case (h), and is in excellent agreement with the approximate ΔSCF result in case (d). Finally, the simple treatment in case (g) is seen to work quite well but could easily be improved by the inclusion of the difference between the *screened* $4\underline{d}$-$4\underline{d}$ and $4\underline{d}$-4f interactions given in Eqs. (67c, d), yielding a shift of ~ 0.2 Ry (without screening ~ 0.4 Ry) and a good agreement with cases (c) and (h).

6.1.2 Interpretation of Results

The development of the experimental XPS spectra in Figs. 1 and 23 can be described essentially in terms of a *discrete level interacting with a continuum with finite band width.* From the results for the self-energy in Fig. 25 one finds that the finite character of the continuum distribution enters in two ways, namely through a more or less sharp cut-off on the low energy side or through condensation of the excitation strength to a narrow range of discrete or quasi-discrete levels. The transition from a broadband to a narrow band situation is directly illustrated in Fig. 29, where the real part of the self-energies in Fig. 25 has been drawn with the $4\underline{p}_{3/2}(\Delta\text{SCF})$ energy as a common reference point.

$\text{Im}\,\Sigma_{\underline{i}}(E)$ can be considered as a product of an ionic excitation density of states and an energy-dependent coupling constant. In model calculations one can independently vary the shape and the band with of the denstiy of states and the strength of the coupling constant. In the present case we can only vary these parameters indirectly by changing the atomic number Z. Since the self-energy involves the polarizability of the ionic system there must be an oscillator-strength sum rule such that

$$\int \text{Im}\,\Sigma_{4\underline{p}}(E)\,dE \propto \text{const.} \propto N_{4d} - 1 \qquad (70)$$

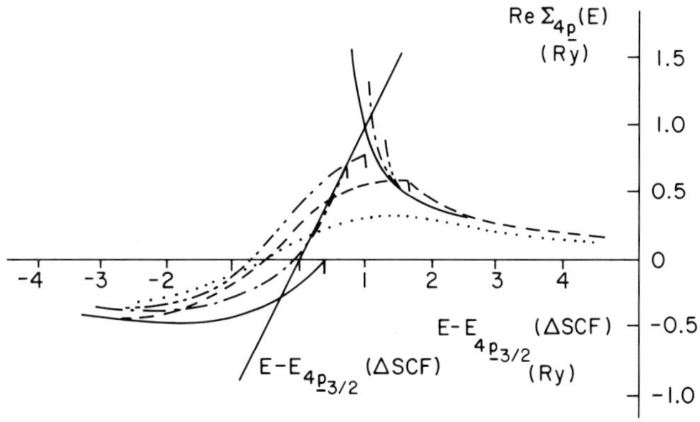

Fig. 29. This figure presents the data in Fig. 25 redrawn with $E-E_{4p_{3/2}}(\Delta SCF)$ as a common reference. It illustrates the transition from a broadband to a narrow band continuum, i.e. from weak dispersion to strong dispersion

From $_{46}$Pd to $_{57}$La the 4 d-shell is completely occupied ($N_{4d} = 10$) and the 4 f-shell empty. The effective number of electrons participating in the excitations is then constant and, provided the 4 p-4 d overlap distribution is approximately constant, the area under the Im $\Sigma_{4p}(E)$ should be approximately constant. The effective coupling strength is then monitored by compressing the constant area continuum distribution to an increasingly narrow band with increasing atomic number Z. In a sense, the coupling strength increases with increasing Z but at the same time the position of the unperturbed discrete levels moves relative to the band. We then obtain the interesting situation that a case of weaker coupling can show more spectacular effects than a case of stronger coupling.

In the transition from a broadband to a narrow band situation, there is a critical case when the positive slope part of Re $\Sigma_{4p}(E)$ becomes parallel to $E-E_{4p}$ (in a point or over a region). On the broad-band side (low Z), we have for *all* E

$$\frac{\partial}{\partial E}\,\mathrm{Re}\,\Sigma_{4p}(E) < 1 \tag{71 a}$$

$$Z_{4p}(E) = \left\{1 - \frac{\partial}{\partial E}\,\mathrm{Re}\,\Sigma_{4p}(E)\right\}^{-1} > 0 \tag{71 b}$$

and there can be only a *single curve crossing* (which need not be any quasi-particle solution). On the narrow band side (high Z) we have, for a *limited range* of E,

$$\frac{\partial}{\partial E}\,\mathrm{Re}\,\Sigma_{4p}(E) > 1 \tag{72 a}$$

$$Z_{4p}(E) < 0 \tag{72 b}$$

and there is now a possibility of having *three crossings*, i.e. three formal solutions to the Dyson equation (Eqs. (15), (58)).

It the unperturbed discrete level is placed inside the region of *anomalous dispersion* where

$$\frac{\partial}{\partial E} \operatorname{Re} \Sigma_{4\underline{p}}(E) > 0 \tag{73}$$

interesting relationships will result. If we consider the case of critical coupling, then E-$E_{4\underline{p}}$ and Re $\Sigma_{4\underline{p}}(E)$ are nearly parallel over an extended energy range and

$$E - E^0_{4\underline{p}} - \operatorname{Re} \Sigma_{4\underline{p}}(E) \simeq 0 \tag{74}$$

in which case the spectral function (Eq. (10)) reduces to

$$A_{4\underline{p}}(E) \quad \simeq \frac{1}{\pi} \{\operatorname{Im} \Sigma_{4\underline{p}}(E)\}^{-1} \tag{75}$$

Typically, the imaginary part of the self-energy peaks in the region of anomalous dispersion. In the spectral function there will then be a *dip* and from this minimum the spectral function will grow until the normal dispersion takes over. Had we started with a Lorentzian continuum distribution, the resulting spectral function would have been a broadened Lorentzian with a pronounced dip at the centre. For weaker coupling we would simply have a Lorentzian with a more or less distorted central portion (corresponding to non-exponential long-time behaviour) while for over-critical coupling the spectral function is split into *two new resonances.*

We now have the necessary background for understanding the behaviour of the XPS spectra in Figs. 24 and 28 through the characteristics of the calculated self-energies in Figs. 25 and 29.

a) $_{46}$Pd: The core level peaks clearly correspond to quasi-particle like solutions but the line shapes are strongly non-Lorentzian. The line width parameter $\Gamma(E)$ (Eq. (19)) varies by 50% over the line width and the high energy parts of the lines lie in the region of anomalous dispersion.

b) $_{48}$Cd: Although there are formally solutions to the Dyson equation, it is probably fair to say that the quasi-particle picture is no longer valid. Only the low energy line wings show Lorentzian-like behaviour. Over most of the line profile the line width parameter varies violently and the high energy side is markedly affected by anomalous dispersion. The spectral function (Fig. 26) becomes strongly asymmetric and no longer shows typical line behaviour.

c) $_{50}$Sn, $_{52}$Te: We are still in the broadband regime but now the line centres have moved into the region of anomalous dispersion. The quasi-particle picture has broken down completely and the spectral distribution is entirely non-Lorentzian, except perhaps for the extreme low energy wings. In a sense, the distributions still look like a very broad and asymmetric line, but it is fairly meaningless to associate the full width at half maximum (FWHM) with any decay width. Since now spectral strength is depleted in the central region of anomalous dispersion and transferred to the wings, there is non-Lorentzian broadening due to the tendency of forming two new resonances by mixing ("CI") the finite bandwidth continuum with the discrete unperturbed core level. The width is therefore beginning to reflect the *splitting* of the spectrum on a number of discrete or quasi-discrete features.

d) $_{53}$I, $_{54}$Xe: The sharp low-energy cut-off is now introducing narrow band characteristics on the low-energy side of the spectrum. The unperturbed discrete levels still lie in the middle of the anomalous dispersion region but, due to the narrow band character of the low-energy side, we are now in the strong coupling regime. As a result, there are formally three solutions to the Dyson equation (Eq. (15)) and a discrete level is split off on the low-energy side. It must be clearly understood that, in principle, this split-off level has nothing to do with the existence of discrete levels below the threshold in the continuum distribution before coupling to the discrete level. It will appear just as well for a pure continuum band, e.g. if the threshold structure has been smeared by a lifetime broadening or solid-state effects. We have actually included dispersion and broadening due to Coster-Kronig processes (Eqs. (63, 64)) but in addition there is a life-time broadening of the $4\underline{d}$ core holes and the $4\underline{d}^2$mf excitations as well as instrumental broadening. All in all, even if the fine structure below the $4\underline{d}^2$ threshold is smeared to a quasi-continuum, a level will be split off from the continuum and give rise to a more or less pronounced peak, below the threshold.

In $_{53}$I a quasi-particle solution is barely emerging and it is probably fair to say that there is still a breakdown of the quasi-particle picture. However, both experimentally and theoretically, there is a pronounced threshold peak which seems to be rather insensitive to the chemical environment. In $_{54}$Xe the split-off peak carries appreciable strength (see Table 2) and clearly corresponds to a *quasi-particle excitation,* while for the remaining part of the spectrum there occurs a complete breakdown of the quasi-particle picture, giving a broad non-resonant continuum. The split-off level originates from mixing of a discrete $4\underline{p}$ level with a band of $4\underline{d}^2\varepsilon$f excitations. In this band the εf wave functions tend to localize resonantly in the region of the 4p and 4d orbitals, thereby acquiring collapsed 4f-like character. The quasi-particle peak then becomes a rather strange object: In the core region it will look like a very compact $4\underline{d}^2$ 4f excitation very strongly perturbed by a $4\underline{p}$ hole, or equivalently, like a rapidly rotating distorted $4\underline{p}$ hole screened by an accompanying tidal wave distortion of the 4d shell. At large radial distances, it will be characterized by the diffuse wave functions of the discrete $4\underline{d}^2$ml structure below the threshold. This is entirely analogous to the case of $4\underline{d}$ 4f excitation in Ba and La[87, 88]. Since the charge density distribution of the excitation is so strongly localized, the spectral distribution becomes rather insensitive to the character of the environment and this is why the present atomic calculations successfully account for a sequence of predominantly solid-state spectra.

It should be noted that as long as one only considers the quasi-stationary split-off level, the problem can be treated using multi-configuration Hartree-Fock techniques. In this way, Aoyagi et al.[89] have calculated a very accurate position for the "$4\underline{p}$" $^2P_{3/2}$ level in Xe.

e) $_{56}$Ba: This is as close as we can come to the strong coupling, narrow-band limit. Most of the ionic excitation strength has now condensed into a collapsed $4\underline{d}^2$ 4f configuration and we are essentially left with a two-level problem. The present results suggest that ~60% of the $4\underline{p}_{3/2}$ hole and ~30% of the $4\underline{p}_{1/2}$ hole remain in the final state as well-defined relaxed core levels. The distribution of the remaining strength we can only speculate about since we have a very poor description of the $4\underline{d}^2$ 4f level structure, as discussed in Sect. 6.1.1. However, it seems reasonable to associate the satellite nearest to the main line with $4\underline{d}(4\underline{d}\ 4f\,^3D,\ ^3P)_{1/2}$ and the next double-peaked satellite with $4\underline{d}(4\underline{d}\ 4f\,^3D,\ ^3P)_{3/2}$. Furthermore, the broad asymmetric peak around the unperturbed $4\underline{p}_{1/2}$ (ΔSCF) position is probably associated with the $4\underline{p}_{1/2}$ level spread over the $4\underline{d}(4\underline{d}\ 4f\,^1P)$ levels and the adjacent continuum, and there is probably also some contribution from the $4\underline{p}_{3/2}$ level. All the spectra from $_{55}$Cs to $_{58}$Ce look very similar and will tentatively be classified as discussed above. Much the same satellite structure seems to persist also in $_{59}$Pr and $_{60}$Nd, resisting the multiplet structure induced by the open 4f shell, simply because the $4\underline{d}$ 4f singlet-triplet splitting remains the strongest source for multiplet splitting.

I would like to add a note of caution regarding the interpretation of the main $4\underline{p}_{3/2}$ XPS line from $_{55}$Cs and upwards. It seems as if the unperturbed level is located right in the middle of the tentative triplet-satellite structure and it is possible that the resulting main line, in a sense, is a very strong $4\underline{d}(4\underline{d}\,4f\,{}^3D, {}^3P)_{3/2}$ satellite line, similar to the case of Xe. However, this is really semantics, since that kind of classification scheme has broken down anyway, and we are really dealing with symmetric and antisymmetric superpositions of two dominant levels or a level and a band.

I would like to finish this section by a discussion of the time dependence of the relaxation process due to gCK fluctuation and decay. The photoionization process suddenly creates a HF-like $4\underline{p}$ core hole and the time development of this hole is very different in the broadband and the narrow-band cases. As discussed in Sect. 3.5, the effect of the gCK fluctuation process is to introduce an angular correlation between dipolar deformations of the charge densities of the $4\underline{p}$ hole and the 4d shell. The $4\underline{p}$ hole begins to create a "tidal wave" in the 4d shell and the time τ_R that it takes to build up this deformed state is the *relaxation time,* given by

$$\tau_R \sim \{\Delta_{4\underline{p}}^{D1}\}^{-1} \tag{76}$$

In order for this relaxed state to be meaningful, the decay time τ_D must be long compared with the relaxation time τ_R, i.e. the line shift must be considerably larger than the line width. In the opposite case, the relaxation process becomes *incomplete.*

In $_{46}$Pd the $4\underline{p}$ width is considerably larger than the shift which means that during the formation of the tidal wave, the energy loss ("friction") is dominant and a maximum relaxation shift is not very probable. However, this does of course not violate the picture of a $4\underline{p}$ hole as a stable excitation in the sense that the total width is still much smaller than the core-hole binding energy. However, in order to observe the Weisskopf-Wigner limit of exponential decay one may have to go back to the elements around $_{42}$Mo.

If we go to the extreme narrow-band limit with two sharp levels, the characteristic time is now the period of the Rabi oscillation between the two states, i.e. there will be a relaxation-charge density oscillation with a frequency given by the resonance integral, i.e. the relaxation shift. If one of the levels is split into a number of quasi-discrete levels (as in $_{56}$Ba) or a continuum (as in $_{54}$Xe), the reversibility will be lost. The $4\underline{p}_{3/2}$ core hole can then start a screening-charge oscillation with a Rabi period time-constant but, due to the dissipation during the relaxation process, the relaxed state will only survive with a certain probability ($\simeq 0.6$ for the $4\underline{p}_{3/2}$ main line in $_{56}$Ba; $\simeq 0.4$ for the main line in $_{54}$Xe). Finally, extreme cases like $_{50}$Sn and $_{52}$Te seem to correspond to overcritically damped behaviour of the screening charge but here the decay is actually so rapid that the core hole has disappeared after typically only a few periods of its orbit.

The problem of a discrete level strongly interacting with a finite bandwidth continuum has a wide range of applications, e.g. to chemisorption and surface problems[50, 90, 91], to the problem of a sharp frequency laser interacting with a broad atomic resonance line[92] and a sharp Rydberg series of atomic absorption lines passing over and interacting with a broad absorption line[93]. In Sect. 8 we shall return to this general problem in connection with solid-state applications.

6.1.3 *The $4\underline{s}$, $4\underline{p}$ X-Ray Photoelectron Spectrum of Xe*

In the case of a single-core hole level the spectral function can directly be compared with experiments (XPS, ESCA). However, in cases with two or more core levels the dipole matrix elements must be included to determine the relative strength of the individual core-level spectra. In the case of spin-orbit split levels, this can be done through statistical weighting of the spectra but for core holes in different shells or subshells, the matrix elements have to be included explicitly. In the $4\underline{s}$, $4\underline{p}$ case one then obtains

$$\frac{dI(\omega)}{d\varepsilon} = C \sum_{m_l,\, m_s} \{|\langle \varepsilon p|z|4s\rangle|^2 A_{4\underline{s}}(E) + (|\langle \varepsilon s|z|4p\rangle|^2 + |\langle \varepsilon d|z|4p\rangle|^2)\, A_{4\underline{p}}(E)\} \tag{77}$$

$$A_{4\underline{p}}(E) = \frac{1}{3}\, A_{4\underline{p}_{1/2}}(E) + A_{4\underline{p}_{3/2}}(E)$$

This result is angle-*independent* and represents an average of the photoelectron current over all angles (electron distribution curve, EDC) or observation at the magic angle. The resulting XPS spectrum for Xe is shown in Fig. 30. The theoretical spectrum has been normalized to the experimental $4\underline{s}$ peak after removal of a small background[19]. This determines the constant C and consequently also the spectrum in the $4\underline{p}$ region. The good qualitative and even quantitative agreement between theory and experiment in Fig. 30 suggests that the previously given characterization of the spectrum is correct. To sum up, the quasi-particle picture for the $4\underline{s}$ hole is perfectly valid but there is a large relaxation shift due to dynamic screening (gCK fluctuations) meaning that the *screening charge tends to follow the motion of the hole.* On the other hand, in the $4\underline{p}$ region the quasi-particle picture has broken down almost entirely. A substantial part (~ 0.4) of the $4\underline{p}_{3/2}$ hole has gone into a kind of quasi-particle excitation in the form of a very compact and strongly distorted $4\underline{d}^2 4f$ configuration, representing the lowest-energy mode of a $^2P_{3/2}$ hole moving in the $n = 4$ main shell. For the remaining larger part of the $4\underline{p}_{3/2}$ core hole strength and for the $4\underline{p}_{1/2}$ core hole there is no quasi-particle character and the spectral strength is smeared to a non-resonant continuum. Finally, after statistical weighting, the "main" line carries only about 25 percent of the total $4\underline{p}$ spectral strength. Regarding the fine structure in the main line we have no explanation except that it probably reflects the growing importance of the $4\underline{d}4f$ singlet-triplet splitting with perhaps some influence from the $4\underline{d}^2$ threshold splitting.

Figure 30 also shows theoretical $4\underline{s}$, $4\underline{p}$ spectra for Pd, Cd and I, compared with experiment. Sn and Te are not shown here since a comparison with experiment is severely complicated by an experimental inelastic background in the $4\underline{p}$ region (Fig. 23) and by incompleteness of the theoretical approximations in the $4\underline{s}$ region. For a further discussion of the $4\underline{s}$, $4\underline{p}$ X-ray photoelectron spectra in the range of elements Pd to Xe and for a more extensive comparison with experiment, we refer to[85].

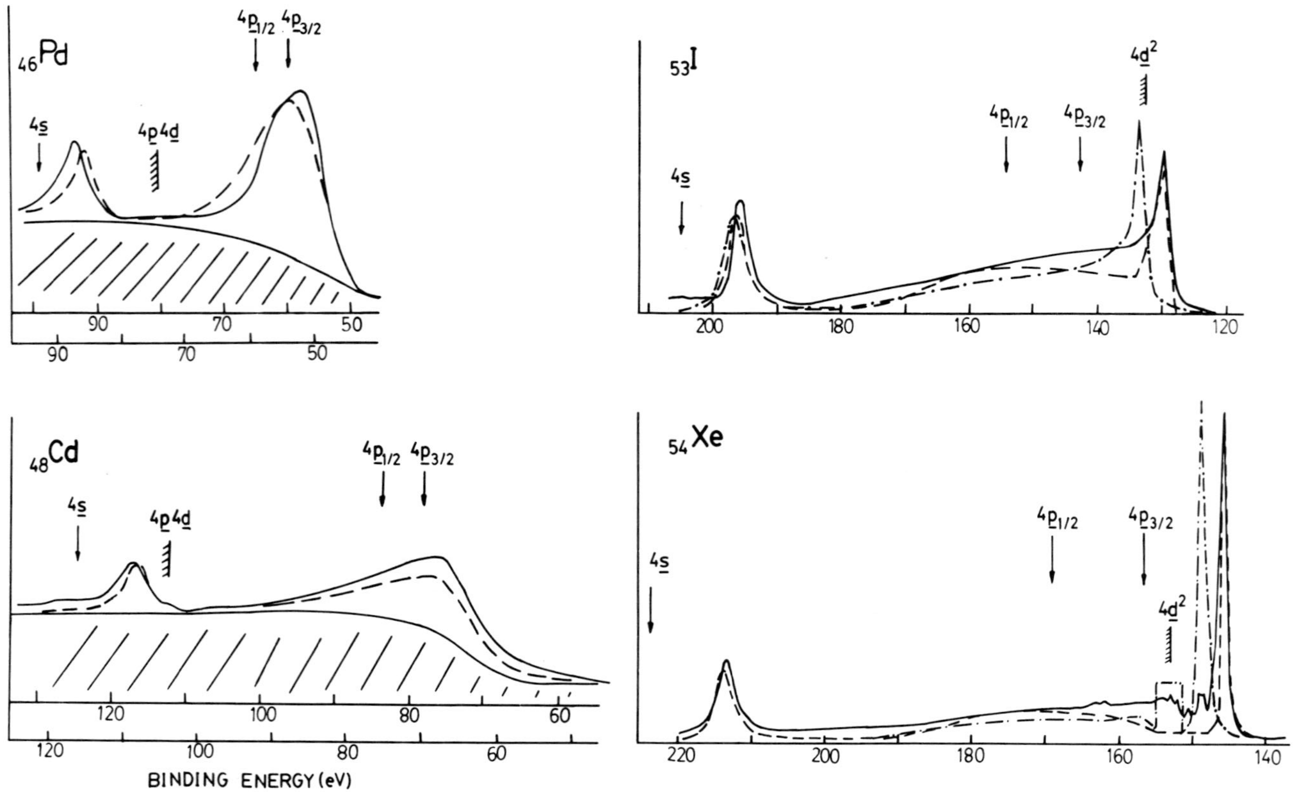

Fig. 30. Experimental and theoretical XPS spectra for Pd, Cd, I and Xe in the 4s, 4p region. Arrows and thresholds refer to relativistic ΔSCF energies. Experimental spectra from[9, 10] (Pd, Cd) and[7, 8] (I, Xe). Lower scale: binding energy for solid

6.2 5 s, 5 p Spectra in the 5f-Series and the Preceding Elements ($80 < Z < 100$)

As already mentioned in Section 4, the 5 s and 5 p core level spectra of the 5f-series and a number of the preceding elements should show pronounced effects of giant Coster-Kronig fluctuation and decay processes much in the same way as happens in the 4 s, 4 p spectra in the range $48 \leqslant Z < 70$. In the case of the 5 p spectrum, the gCK interaction process becomes

$$5\underline{p} \leftrightarrows 5\underline{d}^2 \mathrm{mf} \tag{78}$$

with the interaction strength distribution given to second order by

$$|g^1_m|^2 \simeq \frac{14}{15} R^1(5d5d; 5pmf)^2 \tag{79}$$

(cf. Eq. (54)). The interaction strength should be comparable to the previously discussed 4 p case and one would expect a similar development of the 5 p spectrum with breakdown of the quasi-particle picture. From Fig. 16 one could expect the 5 p levels to be quite broad around $_{83}$Bi and perhaps to disappear in the range $_{85}$At to $_{88}$Ra. After that, a pronounced discrete core-level structure should reappear and there would again be the problem of telling which is the "main" line corresponding to the relaxed core hole and which lines are satellites. In Fig. 16 one notices that, after the level crossing, the $5\underline{d}^2_{av}$ threshold lies closer to the $5\underline{p}_{3/2}$ level than does (Fig. 15) the $4\underline{d}^2_{av}$ threshold to the $4\underline{p}_{3/2}$ level. This implies that the $5\underline{p}_{3/2}$ level might be strongly distorted all the way through the 5f-series.

Unfortunately, there does not seem to exist any compilation of experimental 5 p XPS spectra for the range $80 < Z < 100$ which could tell us directly how the development of the 5 p core-level structure works out in practice. In Fig. 31 we show a few of the available spectra. In atomic $_{80}$Hg (Fig. 31 a) and metallic $_{83}$Bi (Fig. 31b) the $5\underline{p}_{1/2,\,3/3}$ lines are still well-defined but there is a considerable broadening (6 eV) and the levels have become shifted by 3–4 eV to lower binding energy relative to the DF Δ SCF position. Unfortunately, there is a big gap in my collection of data in the very interesting and important range from $_{84}$Po to $_{89}$Ac where the quasi-particle picture can be suspected to gradually break down, eventually obliterating the core levels and leaving a prominent continuum. In view of the large spin-orbit splitting one might also imagine that the $5\underline{p}_{1/2}$ and $5\underline{p}_{3/2}$ levels can be quite differently affected, one level being a good quasi-particle excitation and the other being wiped out.

Figure 31 c shows an XPS spectrum of metallic $_{90}$Th[65]. Comparing the experimental $5\underline{d}_{5/2}$ binding energy with the theoretical value (Δ SCF[82] plus Fermi sea correlation; see Sect. 6.1), we obtain a shift $\Delta^{as}_{5\underline{d}} \simeq 11$ eV between the atom and the solid (Eq. (47)), in quite good agreement with[83]. The energy scale is then shifted by 11 eV to give an atomic scale, assumed to be valid also in the 5 p region. In Fig. 31 c we have marked the positions of approximative $5\underline{p}_{1/2,\,3/2}$ DF ΔSCF binding energies and estimated the position of the various $5\underline{d}^2$ thresholds. In contrast to the analogous cases of Ba to Ce (Fig. 23) it seems likely that the $5\underline{d}^2 5\mathrm{f}$ level structure extends to lower binding energy than the $5\underline{p}_{3/2}$ ΔSCF

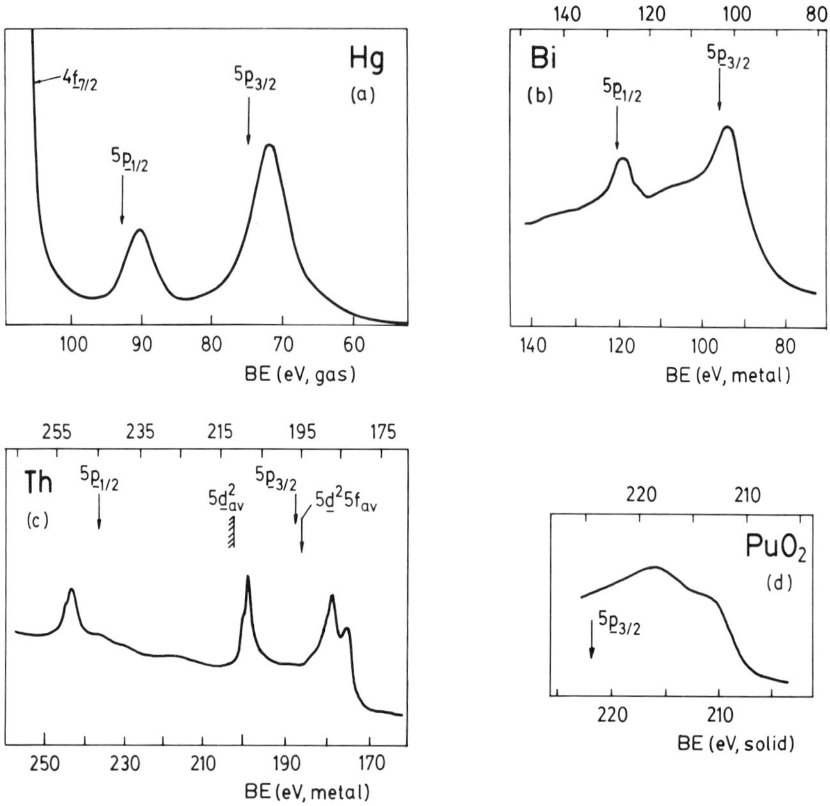

Fig. 31. Experimental X-ray photoelectron spectra for **(a)** atomic Hg[68], **(b)** metallic Bi[10], **(c)** metallic Th[65] and **(d)** PuO_2[61] in the $5\underline{p}$ region, together with estimated atomic binding energy scales (see text), ΔSCF single-[82] and double-hole levels

level, which then becomes surrounded by and strongly coupled to prominent $5\underline{d}^2 5f$ levels. Switching on the gCK fluctuation (dynamic relaxation; CI) process (Eq. 78)) then spreads the strength of the $5\underline{p}_{3/2}$ hole over a wide range of levels[65].

Without any further calculations of the actual zeroth-order ionic excitation spectrum, a more detailed interpretation of the experimental spectrum in Fig. 31 c becomes somewhat speculative. In comparison with the case of Ba (Figs. 23, 28), the present case is complicated by the strong $5\underline{d}$ spin-orbit splitting (~ 7 eV). However, it seems certain that at lower binding energy than the $5\underline{p}_{3/2}$ ΔSCF level there are zeroth-order $5\underline{d}^2 5f$-like ionic excited levels that will pick up considerable spectral strength and probably give rise to the prominent structure around 185 eV on the atomic scale. The problem of interpreting this structure is then analogous to the $4\underline{d}^2 4f$ case in Xe (Sect. 6.1.3). It should be assigned neither to $5\underline{p}_{3/2}$ (as was done in[62]) nor to $5\underline{d}^2 5f$ and might be thought of as the lowest-energy mode of a $j = 3/2$ hole moving around in the $n = 5$ main shell and having no simple projection onto one-electron configurations and spherical harmonics. However, as a guess, I would say that the 185 eV structure (atomic scale) probably is $5\underline{d}^2 5f\ ^2P_{3/2}$-like while the peak at 205 eV corresponds to the excitation that most resembles a $5\underline{p}_{3/2}$ one-electron HF orbital.

Concerning the $5\,p_{1/2}$ level, the DF ΔSCF position lies well above the $5\,d^2$ thresholds in a continuum which should be rather weak since the $5\,d^2 5\,f$ ionic excitations should have exhausted most of the excitation strength. The broadening of the $5\,p_{1/2}$ level is therefore moderate but the shift appears to be quite large, about 3–4 eV, and goes to *higher binding energy*. Strictly speaking, the $5\,p_{1/2}$ line then is a satellite line. In principle, there should also be a balancing spectral strength at lower binding energies, in the 185–205 eV region, associated with e.g. the $5\,d^2 5\,f\,{}^2P_{1/2}$ levels. However, it is difficult to say whether any discrete structure would be observable.

Finally, for $_{94}$Pu in the form of PuO_2[61] only a small part of the 5 p XPS spectrum has been published (Fig. 31 d). In this case, the atom to solid shift is more difficult to estimate since the 5 d peaks are perturbed by multiplet structures, and the value 3 eV obtained is very approximate. Converted to the solid-state energy scale the $5\,p_{3/2}$ ΔSCF[82] binding energy is found to be ~222.5 eV, as shown in Fig. 31 d. We have not investigated this case any further, but judging from the stability of the double structure from Th throughout the actinides, it seems natural to conclude that this double peak derives from $5\,d^2 5\,f\,{}^2P_{3/2}$ and that there is a structure at some 20 eV higher binding energy that more corresponds to a $5\,p_{3/2}$ one-electron like excitation.

For further references and discussion of core-level spectra in the actinides we refer to Kowalczyk[10, 64], Krause and Nestor[63], Sham and Wending[65] and Boring et al.[214].

6.3 5s Core-Hole Spectrum in Xe and Surrounding Elements

The experimental 5 s XPS spectrum in Xe[7, 8] (Fig. 32 a) looks quite similar to the 4 p spectrum in Ba, suggesting that the 5 s level lies just below the $5\,p^2 nl$ level structure, forming a well-defined $5\,s_{1/2}$ quasi-particle excitation and giving considerable strength to a prominent satellite spectrum, mainly $5\,p^2 5\,d$[94, 95]. The basic process is again giant Coster-Kronig fluctuation of the core hole

$$5\,s \leftrightarrows 5\,p^2 5\,d \tag{80}$$

In order to investigate this process we need, as before, the DFS ΔSCF 5 s and $5\,p^2$ energies (Fig. 15), including Fermi-sea correlation. The deviation of the 5 p levels from the ΔSCF positions should be a quite good measure of the correlation energy shift, giving ~0.9 eV. This shift should be approximately valid also for the 5 s hole and, compared with experiment, there is a discrepancy of ~4 eV. Since the monopole relaxation shift is only ~0.8 eV, this is a very large discrepancy, in fact as much as 15 percent of the binding energy.

In the photoabsorption spectrum of Xe the 5 p 5 d excitation does not dominate but in the presence of an extra core hole, the 5 d orbital becomes much more compact and the $5\,p^2 5\,d$ excitation becomes very prominent. As in the 4 p and 5 p cases discussed before, it seems that the $5\,p^2$ threshold multiplet splitting is not the fundamental starting point for describing the $5\,p^2 5\,d$ multiplet structure. Examining the most important two-electron integrals one finds that the $5\,p\,5\,d\,{}^3P$-${}^1D$ splitting and the 5 p spin-orbit splitting both are of the order of 1 eV. The analysis could therefore be based on the 5 p ($5\,p\,5\,d\,{}^3P$, ${}^3D$, ${}^1P$) levels with subsequent further splitting by the other mechanisms.

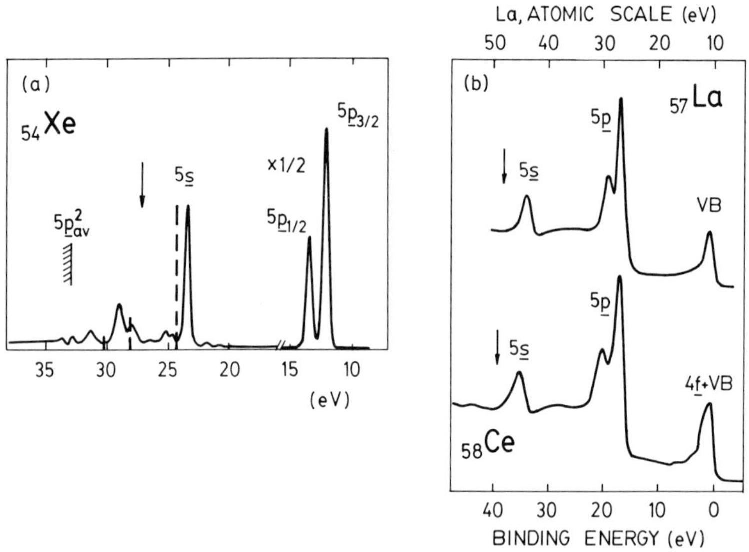

Fig. 32 a, b. Experimental 5s, 5p X-ray photoelectron spectra for **(a)** atomic Xe[7, 8] and **(b)** metallic La and Ce[10]. Positions marked by arrows and thresholds refer to relativistic ΔSCF plus Fermi sea correlation (Δ_i^C chosen to give perfect agreement with the experimental 5p levels). In **(a)**, the vertical dashed lines denote theoretical[94, 95] relative intensities of the 5s, $5p^2 5d$ and $5p^2 6d$ levels

In the actual calculation[94, 95] no attempt was made to include the detailed level structure. The $5p^2md$ ionic excitations were approximated by $5p_{av}$ (5pmd ^1P RPAE) (cf. Sect. 6.1) and the potential was based on the frozen ground-state orbitals. The resulting spectral strength distribution is given in Fig. 32a. The shift of the $5s_{1/2}$ core level of 2.5 eV is of the right order but somewhat too small. Inclusion of the dipole polarizability of the 4d-shell through

$$5s \leftrightarrows 5p\,4d\,mf \tag{81}$$

Coster-Kronig fluctuations increases the $5s_{1/2}$ shift by 0.6 eV to 3.1 eV, showing that there is an appreciable amount of 4d-screening charge that moves along with the $5s_{1/2}$ hole.

Although there is reasonable agreement between theory and experiment, there is still about 1 eV shift to be accounted for and the main line carries too much strength compared with the satellites. However, this can possibly be explained by the fact that the 5p (5p 5d ^3P, ^3D) levels were not included in the calculation and also that ground-state correlations were omitted.

We have chosen to discuss the behaviour of the $5s_{1/2}$ level in Xe in terms of the giant Coster-Kronig fluctuation process in order to underline the analogy with the 4s and 4p core levels. However, since the 5s spectrum is characterized by discrete levels, with rather little weight in the continuum, the problem is of course also suitable for a CI treatment in a more conventional sense. The presence of strong configuration mixing in Xe and in the analogous cases of Ar and Kr has been known for a long time[96] and has been extensively investigated both theoretically and experimentally ([89, 97–103]and references therein). In some recent CI calculations[102, 103] the angular momentum structure of the $5p^4ms$ and $5p^4md$ levels has been treated accurately by resulting in very good descriptions of the spectral strength distributions. The phenomenon has has also been studied in optical excitation spectra of the type $5s^2 5p^5 \rightarrow 5s^1 5p^6$ in the iodine ($_{53}I$) isoelectronic sequence[104–106].

Finally, Figs. 32 b, c demonstrate that in La and Ce metal, the shift of the $5s_{1/2}$ level is at least as large as in Xe, indicating that the giant Coster-Kronig fluctuation process (Eq. (81)) is very strong and atomic-like. However, since the 5 d/6 s electrons now form rather broad bands, one can understand that the satellite structure appears to be washed out in the metal. These types of problems will be further discussed in Section 8.

6.4 5p and 4d Spectra in Atomic Ba. 6s-5d Mixing and Shake-Down

The position of Ba in the periodic system just before the rare earth elements indicates that the electronic charge distribution in the outermost region will be very sensitive to changes in the effective nuclear charge. Previously, in Sect. 6.1, we studied the effect of collapse of the 4f-orbital on the behaviour of 4s and 4p holes. In this section we shall investigate the effect of collapse of the 5d-shell, as already briefly discussed in Section 3.4 (see also[107]).

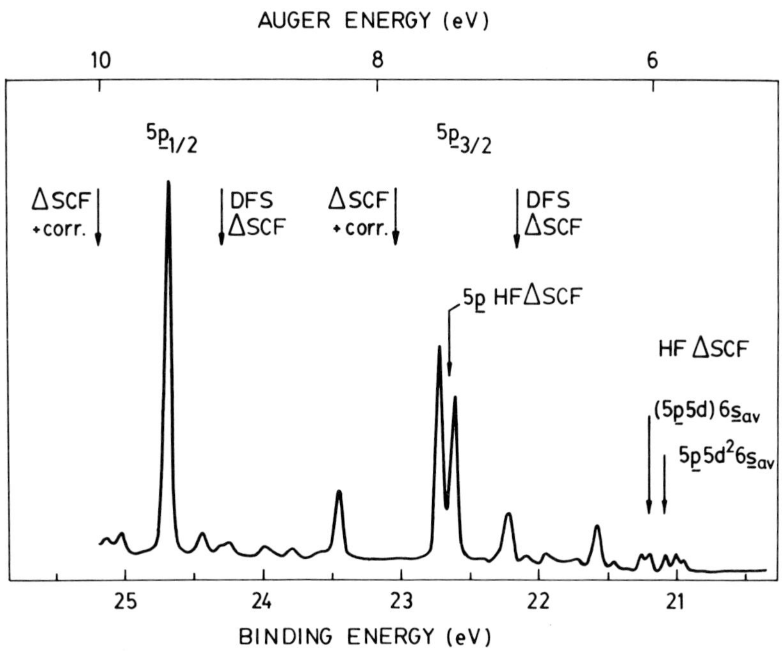

Fig. 33. Ba $5p \rightarrow 6s^2\varepsilon p$ ($O_{23}P_1P_1$) electron-excited atomic Auger spectrum (AES)[54, 55] giving a direct picture of the 5p hole spectrum (see text)

In Fig. 33 we show an experimental O_{23}-P_1P_1 Auger spectrum for atomic Ba by Breuckmann and coworkers[54, 55], which directly mirrors the 5p XPS spectrum since the final ionic state in the Auger process consists of a single level (see also[67, 107]). We can therefore introduce a core-hole energy scale (negative of the binding energy) and consider the spectrum to be an XPS spectrum. In Fig. 33 we have marked the positions of the DFS ΔSCF $5p_{1/2}$ and $5p_{3/2}$ core levels[82] and also the positions obtained by including a shift of 0.9 eV due to Fermi-sea correlations (taken to be the same as in Xe). This type of approach worked very well in the case of Xe[94, 95] but in Ba it turns out to be entirely inadequate. As can be seen in Fig. 33, the main spectral features correlate roughly with the expected positions of the $5p_{1/2,\,3/2}$ core levels but there seems to be a shift of ~ 0.5 eV that is not accounted for. Furthermore, and more significantly, the $5p_{3/2}$ level appears to be split into two levels and in addition there is a prominent and widespread satellite structure.

Evidently, the presence of the 5p core hole causes a radical reorganization of the outermost shells, the attractive Coulomb interaction pulling down unoccupied 5d-levels into the energy and space regions of the occupied ones so that in effect the final ionic state becomes open shell-like (Figs. 34a, b) with a large number of nearly degenerate levels[107]. The non-spherical part of the core hole potential can then induce deformations and even collapse of the outermost 6s-shell by causing transitions to the low-lying 5d levels, (Fig. 34c). The process is perfectly analogous to charge transfer processes in e.g. transition-metal complexes (see Sect. 8), and we refer to it as atomic *intershell charge transfer*. As a consequence of intershell charge transfer, some satellites can move down to lower binding energy than the primary 5p core level (Figs. 34d, e) and we have a situation that is sometimes called "shake-down" in analogy with "shake-up". However,

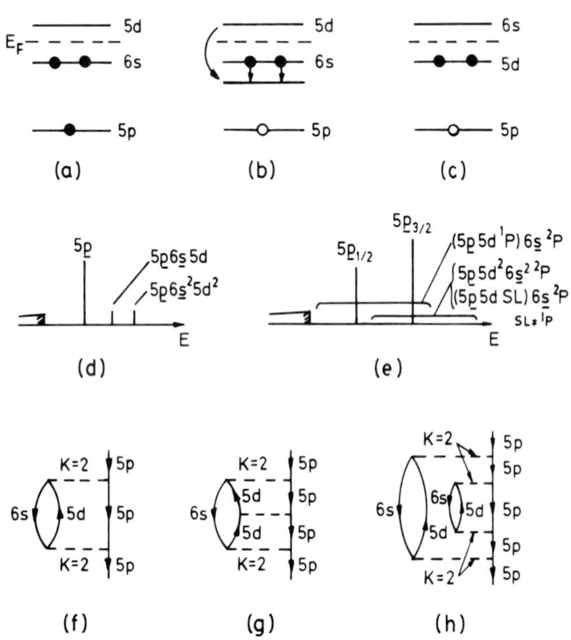

Fig. 34a–h. 5p relaxation processes involving intershell charge transfer ("shake-down"). **(a)–(c)** One-electron level schemes, **(d, e)** schematic 5p hole level spectrum, **(f)–(h)** "shake-down" of one and two 6s-electrons to 5d

it must be realized that the basic shake-down process, in contrast to shake-up, cannot have a monopole character since it involves breaking of the subshell symmetry.

In Figs. 34 f–h are shown some low order 5p self-energy diagrams describing the non-monopole relaxation processes discussed above. Fig. 34 f describes the linear quadrupole (K = 2) response of the 6s-shell where the 5p hole induces the transition (distortion) without disturbing the energy level structure and wave functions. Modification of the level structure enters in lowest order in Fig. 34 g, and included to infinite order it will make the effective 5d-level go below the 6s-level and cause the 5d wave function to collapse into the core region inside the 6s radius. There is also the possibility of transferring two 6s electrons to 5d, leading to the intermediate configuration $5\underline{p}\,6\underline{s}^2\,5d^2$, as shown to lowest order in Fig. 34 h. In this excitation, the effective second-order transition matrix element transfers zero angular momentum to a one-step $6s^2 \rightarrow 5d^2$ double excitation, i.e. we are dealing with a two-electron shake-down or charge-transfer process.

In Fig. 33 we have marked the HF ⊿SCF positions of the $5\underline{p}\,6\underline{s}\,5d$ and $5\underline{p}\,6\underline{s}^2\,5d^2$ levels. However, these levels are of limited value because of the very large exchange and correlation interaction which will spread the levels over a range of about 6 eV. Roughly speaking, we should expect to find the $(5\underline{p}\,5d\,{}^3P)6\underline{s}\,{}^2P_{1/2,\,3/2}$ levels around the $5\underline{p}\,5d\,6\underline{s}_{av}$ position. The $(5\underline{p}\,5d\,{}^1P)6\underline{s}\,{}^2P_{1/2,\,3/2}$ levels would lie at 4–5 eV higher energy but in reality they should go over into a continuum of levels and appear rather weak and diffuse. This highly approximate picture is supported by intermediate coupling calculations by Connerade et al.[52)], who in addition have found a number of important $5\underline{p}\,6\underline{s}^2\,5d^2$ levels spread over the same energy region.

Very recently, Rose et al.[108)] have performed a fully relativistic multi-configuration Dirac-Fock (MCDF) analysis of the problem involving the $5\underline{p}$, $5\underline{p}\,6\underline{s}\,5d$ and $5\underline{p}\,6\underline{s}^2\,5d^2$ configurations. Working essentially in the sudden approximation and calculating the overlap probability between the initial frozen 5p hole and the 5p hole structure in the final relaxed state (cf. Eq. (11) and Sect. 8.3.1), they have found a 5p spectral strength distribution which is in very good qualitative agreement with experiment, as shown in Fig. 35. Their results show that $5\underline{p}\,6\underline{s}^2\,5d^2$ is a very important configuration, meaning that diagrams like Fig. 34 h must not be neglected and actually treated on equal footing with the formally lower-order diagram in Fig. 34 h. The results of Rose et al.[108)] suggest that although the physics can be understood in simple terms, actual calculations of the spectrum may be quite involved. It remains to be seen whether a good qualitative spectrum can be produced using simpler procedures.

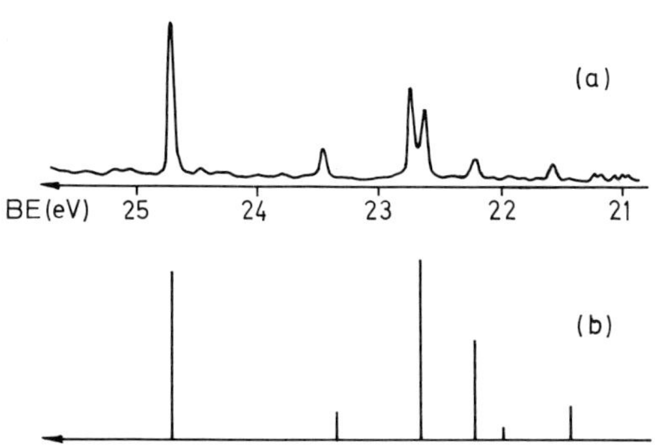

Fig. 35 a, b. Comparison between the experimental 5p spectrum for atomic Ba[55, 56)] (cf. Fig. 33) and the theoretical spectrum of Rose et al.[108)]

The complicated structure of the Ba 5p̲ spectrum is just one example of the effects of inter-shell charge transfer on core-level spectra. Any deep core hole should be able to invert the 6s-5d level structure in Ba and pull empty 5d levels down into the Fermi sea. In order to have actually a symmetry-breaking transition, the core hole must be able to exchange angular momentum, which excludes s-states, and furthermore, the quadrupole-coupling matrix elements must not be too small. The experimental and theoretical results for the 5p̲ level in Ba suggest that the spectrum is strongly disturbed over a region of 5 eV. Then, if the energy resolution is poor, due to the intrinsic width of a deep core level or due to experimental conditions, the spectrum will show no structure, the time for collapse being long compared with the lifetime of the core hole or the characteristic time of the experiment. However, for a number of not too deep core levels the XPS and Auger spectra should show very complicated structures as is clearly demonstrated by the experimental Ba 4d̲ Auger spectrum by Breuckmann and coworkers[54, 55] shown in Fig. 36.

In connection with the N_{45}-P_1P_1 spectrum at around 85 eV, two features are very conspicuous:

a) There are clear satellites on the high-binding energy side of the main $4\underline{d}_{3/2}$ and $4\underline{d}_{5/2}$ lines.
b) The intensity ratio of $4\underline{d}_{5/2}$ to $4\underline{d}_{3/2}$ seems much less than the statistical ratio, indicating that much of the $4\underline{d}_{5/2}$ intensity has become spread over a continuum of levels. Actually, Fig. 36 suggests that the discrete 4d̲ lines are located on a broad continuum bump which may contain the missing discrete spectral strength.

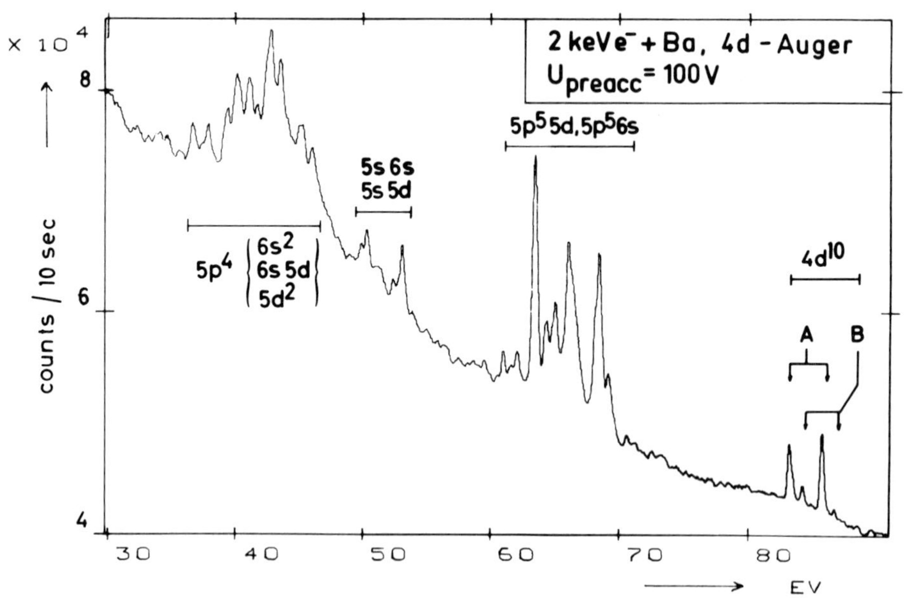

Fig. 36. AES spectrum for atomic Ba[54, 55]. The region between 80 and 90 eV Auger energy directly mirrors the 4d̲ core hole spectrum. In particular, the $4\underline{d}_{5/2}$ level appears strongly perturbed (cf. the $5\underline{p}_{3/2}$ level)

7 Quasi-Particle Properties of Hole Levels in Molecules

In comparison with atoms, molecules can be expected to show a greater variety of many-body effects. This can be understood in several ways. The most direct way is that the outer and inner valence electrons are more or less mobile so that there can be a large build up of screening charge in response to the creation of holes. An equivalent picture is that the single hole level is degenerate or near-degenerate with a number of excited ionic levels with the same symmetry, leading to a strong tendency for the system to reorganize.

In molecules, as well as in atoms, the orbitals are quite stable against relaxation which conserves the symmetry of the orbital, i.e. uniform contraction. However, if the shape of the orbital is allowed to distort in response to the perturbation of a hole, the relaxation energy can become much larger. In the atomic case, large additional relaxation shifts were obtained from correlations between angularly distorted orbitals in which case the screening charge tended to follow the motion of the distorted hole (dynamical screening; Sects. 3 and 6). This effect is of course present in the outer core region of large molecules but in addition, molecules have a new "degree of freedom": In the case of a valence MO (molecular orbital), non-radial distortions tend to localize the charge density to a particular nucleus or group of nuclei, and situations may arise where a localized hole moves together with its equally localized screening charge through the molecule. The symmetry of this quasi-hole is no longer related to the symmetry of the MO's, i.e. the excitation has broken the molecular symmetry.

The question of whether a hole is localized or extended depends in part on the dynamic response of the system. If the static polarizability along the molecular axis is low, one does not gain much relaxation energy by localizing the hole. There is then no way of dynamically distinguishing between a localized and a delocalized hole. Also, even if the static polarizability is high, if the localized wave packet of the hole is constructed from a broad band of levels, the hole can move so fast that the system does not have sufficient time to respond. The screening charge can then no longer follow the hole which therefore appears as unscreened or bare. Since there is no relaxation energy to speak of, the hole is effectively delocalized.

One purpose of this section is to extend the previous discussion on dynamic relaxation of core and valence holes in atoms to molecular systems, first in terms of simple models of the valence level structure and later on in terms of semi-empirical applications to real molecules. A second purpose is to make explicit contact with the recent, closely related theoretical work by Cederbaum and coworkers[109–111)], Bagus and Viinikka[112)] and Martin and Davidson[42)]. For a more complete picture of the previous and recent work in the field, the reader is referred to the above papers[30, 42, 109–112)] and references therein, to the paper by Herman et al.[75)] describing an alternative approach in terms of equations-of-motion, and to the review articles by Shirley[113)] and Basch[114)].

7.1 Simple Model for Dynamic Relaxation

Let us construct a simple model for a molecule by putting together two identical two-level atoms. Each of the atomic levels are then split into a bonding-antibonding pair and we obtain a four-level homonuclear diatomic molecule, as shown in Fig. 37. With later applications in mind, and in order to be very explicit, we assume that the upper levels have π character and the lower levels σ character. We further assume that the π_u level is occupied and the π_g level is empty. We thus have a model with a valence π-bond. The deeper lying σ levels will be assumed to belong to the inner valence or the core region and we shall study the dynamics of the system for a range of band widths W_σ, from $W_\sigma = 0$ corresponding to a deep core level to $W_\sigma \cong W_\pi$ for an inner valence σ level.

We can now discuss what happens when a hole is introduced in one of the occupied σ levels by considering the self-energy and the spectral function for the hole. *Within the present model* there can be no monopole relaxation in response to the removal of an MO because we have eliminated all unoccupied levels with σ_u, π_u or σ_g symmetry. There is then no mechanism for building up screening charges by shape-conserving radial contraction of the molecular charge density, and a ΔSCF calculation would not result in any relaxation shift of the energy of the hole. The model is instead designed to break the inversion symmetry of the individual MO's by introducing a polarizability of the σ- and π-bonds along the molecular axis. By fluctuating between the $\underline{\sigma}_g$ and $\underline{\sigma}_u$ levels, the hole can induce virtual transitions between the π_u and π_g levels (Figs. 38 a, b). This means that the hole no longer has pure g or u symmetry and consequently has a degree of localization on either nucleus, the localized part of the hole oscillating between the nuclei with angular frequency $\omega = W_\sigma/\hbar$. Likewise, the $\pi_u \to \pi_g$ virtual transition represents a localization of the π-bond charge towards either nucleus and we now have a mechanism for π-charge to become displaced along the molecular axis in order to screen a hole that tends to localize on a particular nucleus.

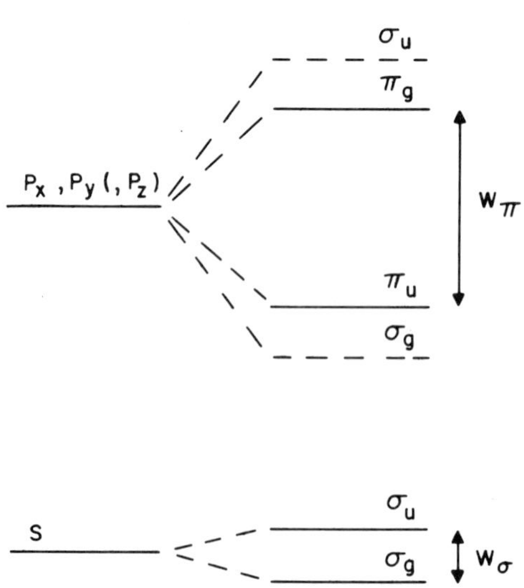

Fig. 37. Schematic picture of the formation of σ and π levels for a homonuclear diatomic molecule

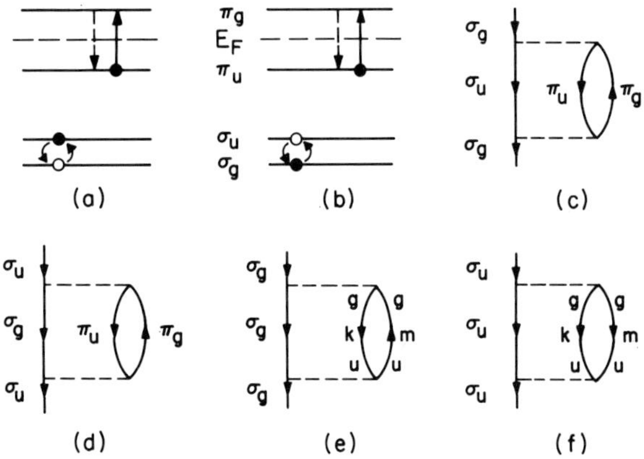

Fig. 38 a–f. One-electron level pictures and self-energy diagrams for fluctuation and relaxation of hole levels within a molecular model-level structure (for explanation of (a) to (f) see text)

The fluctuation process is shown in Figs. 38 c, d in the form of self-energy diagrams for a $\underline{\sigma}_g$ hole (Fig. 38 c) and a $\underline{\sigma}_u$ hole (Fig. 38 d). Since the σ_g and σ_u MO's are situated in the same region of space, we are dealing with a *molecular version of the Coster-Kronig process* discussed previously. Furthermore, if the σ levels also belong to the valence region, then the fluctuation process takes place entirely within the σ-π valence charge region. We then have a *molecular giant Coster-Kronig fluctuation process,* resulting in very large shifts of the unperturbed energy levels.

As already mentioned before, if one could localize a core hole on a given nucleus it would oscillate with a hopping frequency determined by the level splitting W_σ, and the π-charge would have a relax in the presence of a moving hole, i.e. *dynamic relaxation.* However, for a deep core hole the hopping frequency would be practically zero and we would then have *static relaxation* (cf. Sect. 3). We shall now discuss these two cases in some detail.

(a) *Deep core hole, static relaxation.* Since the $\underline{\sigma}_g$ and $\underline{\sigma}_u$ levels are nearly degenerate, the energy transfer ω to the π-electron system is nearly zero (typically of the order of meV) and therefore the core hole is *statically* screened. The only difference between the $\underline{\sigma}_g \leftrightarrows \underline{\sigma}_u$ fluctuation process (Fig. 38 c) and the $\underline{\sigma}_g \leftrightarrows \underline{\sigma}_g$ static monopole relaxation process (Fig. 38 e) is then the matrix element coupling the hole to the rest of the system. Since the molecular orbitals can be built up by linear combination of non-overlapping atomic core orbitals

$$\phi_g = \frac{1}{\sqrt{2}} (\phi_1 + \phi_2) \tag{82 a}$$

$$\phi_u = \frac{1}{\sqrt{2}} (\phi_1 - \phi_2) \tag{82 b}$$

we immediately obtain, using $\phi_1^* \cdot \phi_2 = \phi_2^* \cdot \phi_1 = 0$ (see also[41])

$$|\phi_g|^2 = |\phi_u|^2 = \frac{1}{2} (|\phi_1|^2 + |\phi_2|^2) \tag{83 a}$$

$$\phi_u^* \cdot \phi_g = \frac{1}{2} (|\phi_1|^2 - |\phi_2|^2) \tag{83 b}$$

The static monopole relaxation process (Fig. 38 e) involves the matrix element

$$\langle \sigma_g k | 1/r_{12} | \sigma_g m \rangle = \int |\phi_{\sigma_g}(\mathbf{r})|^2 \frac{1}{|\mathbf{r}-\mathbf{r}'|} \phi_k^*(\mathbf{r}')\phi_m(\mathbf{r}') d\mathbf{r} d\mathbf{r}' \tag{84}$$

and represents the response to a charge distribution

$$|\phi_g|^2 = \frac{1}{2} (|\phi_1|^2 + |\phi_2|^2)$$

where the core hole has the same weight on all the nuclei. The fluctuation process (Fig. 38 c) involves the matrix element

$$\langle \sigma_g \pi_u | 1/r_{12} | \sigma_u \pi_g \rangle = \int \phi_{\sigma_g}^*(\mathbf{r}) \phi_{\sigma_u}(\mathbf{r}) \frac{1}{|\mathbf{r}-\mathbf{r}'|} \phi_{\pi_u}^*(\mathbf{r}') \phi_{\pi_g}(\mathbf{r}') d\mathbf{r} d\mathbf{r}' \tag{85}$$

and represents the response to a *displacement of charge*

$$\phi_{\sigma_u}^* \cdot \phi_{\sigma_g} = \frac{1}{2} (|\phi_1|^2 - |\phi_2|^2) .$$

The static monopole relaxation diagram clearly describes the response to a *delocalized* core hole while the dynamic relaxation (fluctuation) process describes the response to the dipole moment of the hole. Together, the two diagrams (Figs. 38 c, e) describe the response of the system to a hole localized to an atomic core orbital on either nucleus[41)], and individually they represent a multipole expansion of the hole.

For a core hole level in a homonuclear diatomic molecule the radial and axial relaxation energies are about of the same size[40)]. However, in contrast to the axial dipole relaxation, the radial monopole relaxation cannot be described quantitatively within a basis limited to the valence electrons. For a proper description of the radial contraction of an MO one must involve highly excited discrete and continuum orbitals. Therefore, the calculation of the relaxation shift of a delocalized hole requires a large basis set while a description of the localization of the hole can be given within a more limited basis.

b) *Inner valence hole, dynamic relaxation.* We now consider the case that the $\underline{\sigma}_g$, $\underline{\sigma}_u$ levels are no longer degenerate, a situation which is typical of inner valence levels[115)]. In comparison with the static case, the major difference is that the fluctuation self-energy diagram in Fig. 38 c now involves the *dynamic response* of the π-electron bond at the frequency $\omega = W_\sigma/\hbar$. In a real space and time picture this means that the dipolar charge distribution now oscillates with frequency ω, corresponding to a localized hole which hops between the nuclei with frequency ω. As long as $\omega < W_\pi/\hbar$ the π-charge will respond in-phase and follow the hopping hole, i.e. the bare hole and its screening π-charge cloud, *the quasi-hole,* will propagate as one unit. The effect of the fluctuation (correlation) process in Fig. 38 c is therefore to cause additional relaxation by breaking the molecular symmetry and localizing the hole. This kind of correlation process is usually called *dynamic relaxation.*

When the hopping frequency of the $\underline{\sigma}$-hole is equal to the resonance frequency of the π-bond there are two equally important normal modes of the total system $\underline{\sigma}$-hole and π-bond, one where the *hole and screening charge oscillate together in phase* between the nuclei and one where they *oscillate out-of-phase,* the hole and the screening charge being located on different nuclei. This last mode where the hole and its screening charge oscillate relative to each other is closely related to the so-called plasmaron excitation proposed by B. Lundqvist[116] in the case of an electron gas. For $\omega < W_\pi/\hbar$ the in-phase solution will have largest intensity and we have a dynamically screened quasi-hole as the most probable excitation. For $\omega > W_\pi/\hbar$, the out-of-phase mode will dominate. However, there is then a very weak build-up of screening charge and the hole will no longer be *dynamically* screened. One can also say that the hole oscillates so fast that from the point of view of the π-electrons, it appears as delocalized. Left is only the *static* relaxation due to the removal of the charge distribution of an MO extending over the entire molecule.

7.2 Self-Energy and Spectral Function for a Realistic Model of Multiple-Bonded Molecules

By extending the space of the simple model to all valence MO's generated from 2s and 2p electrons in e.g. carbon or nitrogen, we obtain a model which illustrates some fundamental properties of multiple-bonded molecules containing atoms like carbon, nitrogen and surrounding elements. More generally, the model should apply to molecules built from open shell atoms with valence configurations nsnp all the way through the periodic table.

In the simplest type of LCAO model the molecular valence-level structure will look like shown in Fig. 39a. Since the atomic 2s and 2p orbitals have the same spatial extension, the splitting of the levels will be approximately equal, and there is then a fundamental degeneracy. The fluctuation of a $2\underline{\sigma}_g$ hole to $2\underline{\sigma}_u$ will then be nearly resonant with $1\pi_u \rightarrow 1\pi_g$ and $3\sigma_g \rightarrow 3\sigma_u$ excitations, and the $2\underline{\sigma}_g$ hole becomes strongly coupled to the 1π, 3σ bonding valence charge, as discussed in Sect. 7.1. Typical systems having this necessary $2s\sigma$-$2p\pi(2p\sigma)$ multiple bonded structure are the *unsaturated* hydrocarbons like C_2H_2, C_2H_4 and systems like N_2 and O_2. Although the details of the

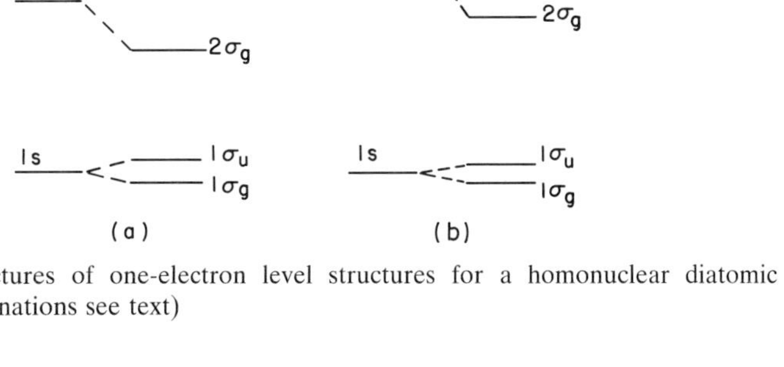

Fig. 39a, b. Schematic pictures of one-electron level structures for a homonuclear diatomic molecule (for further explanations see text)

degeneracy vary between different molecules, the general feature is typical, and one can quite generally expect an MO picture to break down completely in the inner valence region of these molecules (see also[117–119]).

In order to make detailed predictions one must accurately know the dynamic axial dipole polarizability of the outer valence σ-π bonds in the presence of an extra valence hole. In the simplified model above, we essentially used the virtual HF levels for excited states. In HF calculations of molecular orbitals one typically obtains level structures like the one shown in Fig. 38 b, with the $1\pi_g$ virtual level a few volts above the vacuum level. In the presence of the Coulomb potential of a $1\underline{\pi}_u$ hole, the $1\pi_g$ electron becomes quite strongly bound but one has then also to consider the coupling of electron-hole pairs as described by the RPA. The coupling is so strong that the axial dipole response becomes highly collective. The singlet-triplet splitting is a good measure of the collective shift and in the systems considered, it is often $\sim 50\%$ of the triplet excitation energy[120]. The singlet excitation is therefore highly collective and corresponds to a "plasmon" excitation in the bonding charge. The $1\pi_u \rightarrow 1\pi_g$ excitation level structure can then be like shown in Fig. 39 b. Furthermore, in the presence of an additional valence hole, the excitation energy will increase somewhat and, as a result, the singlet level in the ion will not be far away from the virtual HF level in the neutral molecule. Therefore, more accurate calculations do not seriously modify the conclusions based on the simplified model above (Fig. 39 a).

If we take the HF MOSCF occupied levels as a zeroth-order approximation, the hole levels are then influenced by a number of relaxation and correlation effects.

a) *Static monopole relaxation* (Fig. 40 a). These processes are essentially outside the model space. The relaxation shift for the main line can be found from a MOΔSCF calculation. Typical shifts for the valence levels are ~1–1.5 eV to lower binding energy.

b) *Ground-state correlation* (Fig. 40 b). Describes the loss of correlation energy when a number of correlated electron pairs involving the removed core electron are broken up. Typical shifts are ~1.5 eV to higher binding energy[30]. These processes are also largely outside the model space.

Clearly, the monopole relaxation and ground-state correlation shifts approximately cancel, and the HF MO eigenvalues should lead to reasonable binding energies for the valence levels provided there were no other important interaction mechanisms.

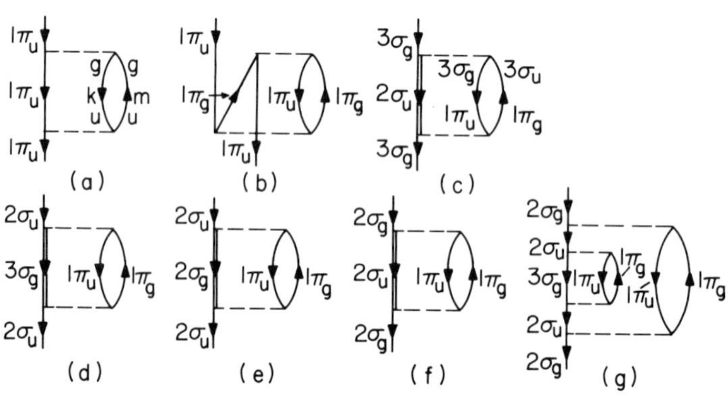

Fig. 40 a–g. Self-energy diagrams describing relaxation and correlation in the valence region of the model level structure in Fig. 39 (for further explanations see text)

c) *Dynamic dipole relaxation* (Figs. 40 c–g). Quite often, fluctuation of the hole to other levels plays an important role. Of all the valence levels in Fig. 39, only for the $1\underline{\pi}_u$ level there is no fluctuation process of any importance and the $1\underline{\pi}_u$ level therefore remains approximately at the HF MO position. The other hole levels are strongly influenced by fluctuation (dynamic localization) through the processes shown in Figs. 40 c–f. An important point to be noted is that the intermediate level to which the hole can fluctuate is also dressed by a self-energy. As a concrete example, consider the dynamic self-energy for the $2\underline{\sigma}_g$ hole in Fig. 40 f. Through its own dynamic self-energy in Fig. 40 b the $2\underline{\sigma}_u$ hole will be dynamically shifted by ~2 eV to lower binding energy and it will also acquire a satellite spectrum based on $3\underline{\sigma}_g 1\underline{\pi}_u 1\pi_g$ ionic excited levels. Therefore, in the intermediate level of the $2\underline{\sigma}_g$ self-energy diagram (Fig. 40 f) all excited ionic states will be based on the dynamically shifted $2\underline{\sigma}_u$ level or on its satellites (Fig. 40 g). As a result, the ionic excited levels based on the $2\underline{\sigma}_u$ level will be shifted by ~2 eV relative to the zero-order $2\underline{\sigma}_g$ level and in addition there will be a spectrum of *ionic double excitations* like $3\underline{\sigma}_g 1\underline{\pi}_u 1\pi_g$ in the $2\underline{\sigma}_g$ region. This means that the relaxation of a particular hole level often cannot be treated as isolated from the relaxation of other levels. There can be a strong coupling so that the entire hole spectra of several primary hole states have to be solved for in a self-consistent manner. Admittedly, this is quite elementary and well-known in the formal many-body theory and in the theory of extended many-particle systems but, to my knowledge, it still awaits serious application in the many-body theory of atomic and molecular systems. For instance, Cederbaum et al.[30, 109–111, 117–119] only calculate the ionic excited levels within the 2 ph TDAE or RPAE, including the bare ladder and bubble interactions among the two holes and the excited electron in the intermediate state of the self-energy to infinite order (Figs. 18 c–f, r, s). However, such an approach neglects relaxation and screening effects in the ionic excited states (Figs. 18 g–m) as well as double and higher excitations associated with shake-up and Auger processes in the final ionic state. Therefore, even if a large basis set makes the resulting hole level spectrum look very detailed and complicated, one has to be somewhat cautious when assigning a particular configuration or state to a particular experimental peak[121]. For instance, Bagus and Viinikka[112], in a CI calculation for the CO molecule limited to the type of model space discussed here, found doubly excited ionic states to be very important for describing the level structure in the $3\underline{\sigma}$ region (corresponding to the $2\underline{\sigma}_g$ region of the isoelectronic molecule N_2).

7.2.1 A Model Spectrum for the $1\underline{s}$ Core Level

Before we proceed to discuss the valence spectrum in greater detail, it is very instructive first to consider the $1\underline{s}$ deep core level spectrum, which is not complicated by the presence of several overlapping satellite spectra and therefore gives direct information about the ionic excitation level structure. If the core level spectrum is experimentally available, this may then be of great help for interpreting valence level spectra.

In the case of a deep core level, the photoelectron will leave the core region in a time short compared with the time for the localized $1\underline{s}$ hole to hop between the nuclei, and the deep core hole will in practice be localized on one of the nuclei. Diagrammatically this is shown in Fig. 41, where Figs. 41 a–d describe photoionization from the symmetry adapted $1\sigma_g$ and $1\sigma_u$ MO's and where Fig. 41 e describes the interference between the

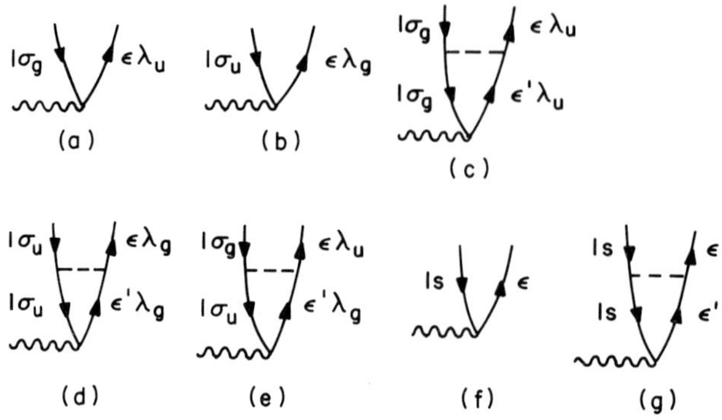

Fig. 41 a–g. Low-order diagrams describing photoionization from delocalized and localized core levels (for further explanations see text)

nearly degenerate $1\sigma_g$ and $1\sigma_u$ levels. As long as the photoelectron energy is much larger than the level splitting, the photoelectron will only sample the charge distribution of the average level (this is generally true and includes photoionization from multiplet levels in open shell atoms and molecules), which is a $1\underline{s}$ hole (cf. Eq. (82)) localized on either nucleus, as shown in Figs. 41 f, g (cf.[122]). Note however, that this *does not mean* that we are describing the molecular photoionization process as the sum of the free atom processes. On the contrary, the description includes in principle a proper account of the threshold behaviour with localized excitons and other prominent structures as well as interference effects like EXAFS.

Figures 42 a–c show some fundamental core level self-energy diagrams, extending the description in Fig. 38 and Sect. 7.1 to a larger valence level model space (for simplicity, we only consider the $1\underline{\sigma}_g$ self-energy). Figures 42 a–c describe the ionic excitation spectrum in terms of virtual HF orbitals while the diagrams in Figs. 42 d–f represent lowest order electrostatic corrections due to the core and valence holes. The actual influence of the holes depends very much on whether the ionic excitation is compact (localized) or diffuse. Strongly localized electron-hole excitations like $1\pi_u \rightarrow 1\pi_g$ are only weakly perturbed by the presence of a second hole and a reasonable zeroth-order approximation will then be simply to add the molecular electron-hole excitation energy to the binding energy of the second hole, the diagrams in Figs. 42 d, e cancelling each other and not giving rise to any net energy shift. Knowledge of the dominating $\pi \rightarrow \pi^*$ dipole transitions (as observed for instance in photoabsorption) would then immediately give a rough idea about the low-lying satellite structure in the core level spectrum.

Let us illustrate these ideas by quoting some numerical results for the N_2 molecule. In the virtual HF orbital scheme, the $1\pi_u \rightarrow 1\pi_g$ transition energy is ~20 eV[120] but taking into account the average Coulomb attraction of the $1\pi_u$ valence hole reduces the excitation energy to ~9 eV[120]. This gives a $1\pi_u$-$1\pi_g$ screened average Coulomb interaction of ~11 eV and also represents an estimate of the effective Coulomb repulsion among valence holes as well as between a core hole and a valence hole. Furthermore, the $(1\underline{\pi}_u 1\pi_g)_{av}$ level represents a one-electron MO picture where the π-electrons respond individually to external perturbations. For singlet excitations this is a very poor picture because the very strong axial polarizability of the π-bond will couple the individual excitations to a collective excitation of the entire π-bond. The dipole-dipole interaction between the $(1\underline{\pi}_u 1\pi_g)_{av}$ electron-hole pairs enters to lowest order through the diagram in

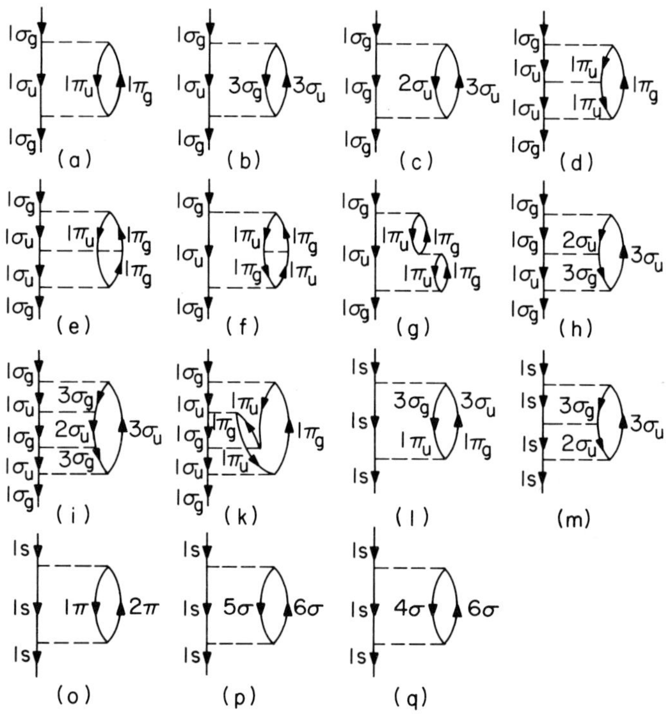

Fig. 42 a–g. Self-energy diagrams for $1\underline{\sigma}_g$ and $1\underline{\sigma}_u$ core holes (see text)

Fig. 42 g, which is one of the basic ingredients in the RPAE (random-phase approximation with exchange; see Sect. 5 and[14, 24, 25]), and leads to a collective shift of the $^1(1\underline{\pi}_u 1\pi_g)$ level by ~8 eV to higher energy[120]. An equivalent approach is of course to think in terms of multiplet structure of the $(1\underline{\pi}_u 1\pi_g)$ level, modified by Fermi sea correlations, the singlet-triplet splitting being determined essentially by the exchange type of matrix element $\langle 1\pi_u 1\pi_g | 1/r_{12} | 1\pi_g 1\pi_u \rangle$ (i.e. the bubble interaction in Fig. 42 g). The calculation by Rose et al.[120] also goes beyond the RPAE and includes relaxation effects and the final results for the $1\underline{\pi}_u 1\pi_g$ triplet and singlet levels are 8 eV and 15 eV, in good agreement with optical spectra.

Comparison with experimental XPS data[7], to be further discussed in Sect. 7.3, suggests that the effect of the deep core hole on the $\pi \to \pi^*$ transition is marginal, only increasing the excitation energy by ~1 eV. Therefore, when calculating the $\pi \to \pi^*$ satellite excitation energies, it seems to make little difference whether one considers the deep level to be localized or delocalized. However, this need not necessarily be true of the entire satellite spectrum: Since localization breaks the gerade-ungerade symmetry of the final state there need not always be any clear distinction between monopole shake-up and g-u dipole shake-up. At present, we do not know to what extent symmetry breaking influences the positions and intensities of the satellites and we shall only briefly indicate the way these effects enter in the diagrammatic expansion.

In Figs. 42 b, c the ionic excited states have the full molecular symmetry and by including diagonal hole-hole and hole-electron ladders in analogy with Figs. 42 d, e will not change this fact, the molecular excitation being influenced by a delocalized $1\underline{\sigma}_g$ or $1\underline{\sigma}_u$ core hole. As before, localization enters through scattering (fluctuation) of the core hole between $1\underline{\sigma}_g$ and $1\underline{\sigma}_u$ levels (Figs. 42 h–k), which can be translated into self-energy diagrams involving localized $1\underline{s}$ core holes (Figs. 42 l, m). The sum of the infinite series of ladder diagrams like Figs. 42 h–k or 42 l, m mixes the monopole excitations (Fig. 42 c) and the g-u dipole excitations (Fig. 42 b) and represents a description of the ionic excitations in the non-symmetric potential of a localized $1\underline{s}$ core hole. This will then distort the g-u symmetry orbitals and the self-energy diagrams involving proper final state orbitals will look like shown in Figs. 42 o–q.

A schematic picture of the real part of the $1\underline{\sigma}_g$ ($1\underline{\sigma}_u$) core hole self-energy is shown in Fig. 43 a and the corresponding spectral function is given in Fig. 43 b. An important point to be noted is that although the low-lying excitations give rise to a pronounced shake-up structure, they only contribute a minor part (about one third) of the total core relaxation shift. The larger part of the relaxation shift is determined by the ionic excitation *continuum,* which is responsible for the formation of the screening charge in the

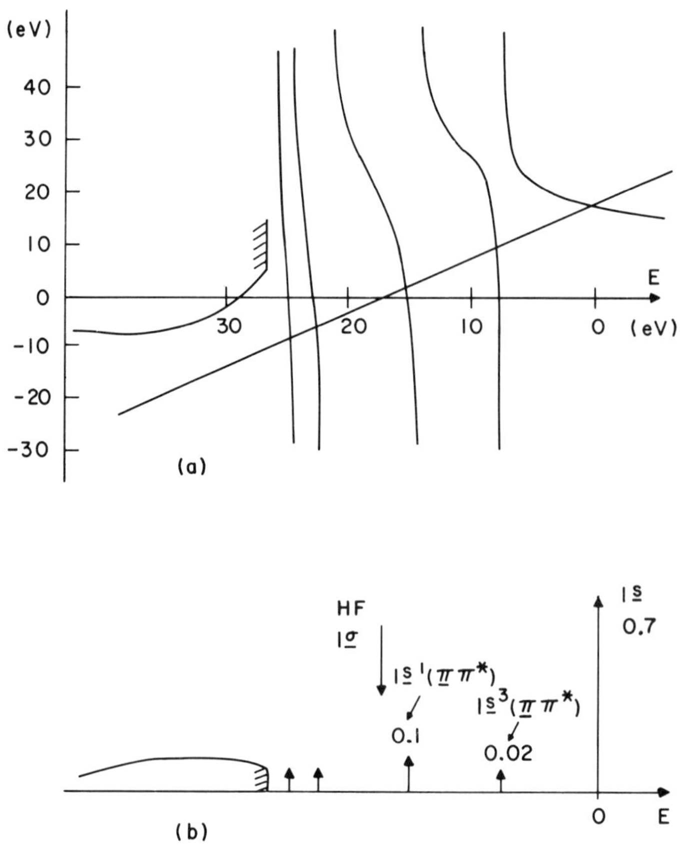

Fig. 43 a, b. Schematic pictures of **(a)** the real part of the self-energy for a $1\underline{\sigma}_g$ (or $1\underline{\sigma}_u$) core hole and **(b)** the corresponding spectral function

close neighbourhood of the hole where the Coulomb potential is very strong. Since the compact $1\underline{s}\,(\underline{\pi}\pi^*)$ ionic excitations have approximately the same relaxation shifts as the parent $1\underline{s}$ level itself, they will move below the HF MO $1\underline{s}$ position and the centre-of-gravity will therefore not lie between the main line and the first shake-up satellite line, as is usually the case with atoms.

Since the $\pi \rightarrow \pi^*$ excitation can be thought of as plasmon-like collective excitation of the π-bond, one could also imagine the core hole to shake up two plasmons with approximately additive energies. As we shall see later, these excitations fall in a range where there is a large amount of shake-up intensity and it is just possible that they play a certain role.

7.2.2 Model Spectra for Valence Hole Levels

From the previous discussion we have a reasonable idea about the positions of the $\underline{\pi}\pi^*$ excitations relative to the ionic ground state, and we now simply transfer this picture to the valence hole levels. Fig. 44 a shows a typical, schematic picture of the self-energy for

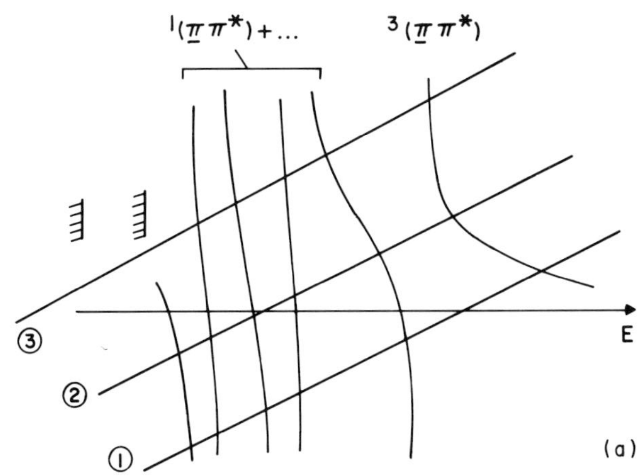

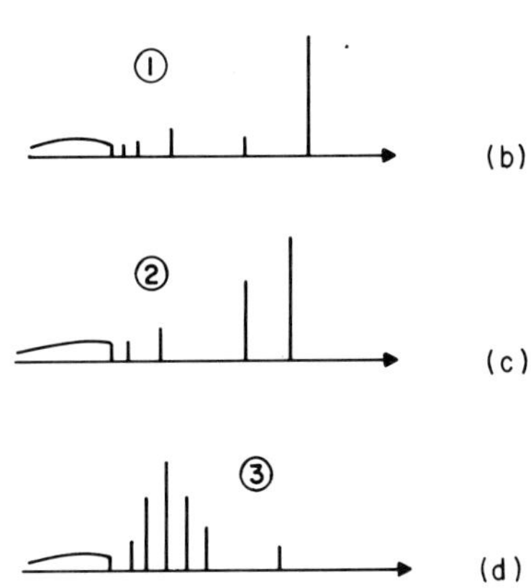

Fig. 44. **(a)** Schematic picture of the real part of the self-energy for a valence hole, together with graphical solutions ①, ② and ③ of the Dyson equation (Eq. (15)). **(b)** is meant to represent a typical outer-valence hole spectrum while **(c)** and **(d)** describe the possible behaviour of inner-valence holes. **(b)**–**(d)** are connected with the solutions ①–③ resp. Note that in principle the self-energy is different for different valence holes, contrary to what is suggested in **(a)**

a valence hole that has a possibility of fluctuating among valence levels (Figs. 40 c–f). A number of solutions of general interest of the Dyson equation (Eq. (15)) have been marked and the corresponding valence hole spectra are shown in Figs. 44 b–d. Fig. 44 b shows a typical spectrum for outer valence holes like $3\underline{\sigma}_g$ and $2\underline{\sigma}_u$. This spectrum is quite similar to the $1\underline{s}$ core level spectrum and represents the shake-up in response to a slowly hopping hole instead of a static, localized one. Fig. 44 c describes the situation that the hopping frequency of the valence hole is nearly the same as the ionic $^3(\underline{\pi\pi}^*)$ excitation energy in which case the hole level is split into two levels of comparable intensity. We shall later see (Sects. 7.3.5–6) that this situation is typical of the $2\underline{\sigma}_g$ inner valence level in hydrocarbons like C_2H_2 and C_2H_4. Finally, Fig. 44 d describes a situation where the hopping frequency of the hole is nearly equal to the energy of the $^1(\underline{\pi\pi}^*)$ ionic excitations, among others, the hole level becoming distributed over several intense components and also giving rise to important satellite structures on the low-binding energy side.

The valence hole model spectra shown in Figs. 44 b–d should give a rough idea about what kind of hole structure one can expect to find in real systems like N_2, O_2, CO, N_2O, C_2H_2, C_2H_4 etc. Taking N_2 and C_2H_2 as specific examples, all levels except $1\underline{\pi}_u$ are strongly affected by dynamical dipole relaxations. The outer valence $3\underline{\sigma}_g$ and $2\underline{\sigma}_u$ spectra behave essentially as the $1\underline{\sigma}_g$, $1\underline{\sigma}_u$ levels with a relaxed main line and several satellites at higher binding energy, while the $1\underline{\pi}_u$ level essentially does not couple to the g-u dipole excitations and therefore remains delocalized. Depending on the system (i.e. the positions of the HF molecular orbital energies), the $3\underline{\sigma}_g$ level may or may not move past the $1\underline{\pi}_u$ level and one may have the HF MO ordering $1\underline{\pi}_u$, $3\underline{\sigma}_g$, $2\underline{\sigma}_u$ as in C_2H_2 or the inverted ordering $3\underline{\sigma}_g$, $1\underline{\pi}_u$, $2\underline{\sigma}_u$ as in N_2. However, in the case of the inner valence $2\underline{\sigma}_g$ level, the very shape of the spectrum is strongly system-dependent. The recent calculations by Cederbaum and coworkers show that the C_2H_2 and C_2H_4 $2\underline{\sigma}_g$ spectra[111)] correspond to Fig. 44 c which, in a sense, represents a normal spectrum[111)] while systems like N_2, O_2 and F_2[109, 123)] correspond to Fig. 44 d, describing a strong coupling case where the $2\underline{\sigma}_g$ level becomes spread over a wide range of ionic excitations. One can argue that this represents a breakdown of not only the MO picture but also of the quasi-particle picture (cf. Sect. 7.3.7) for an inner valence hole. At any rate, the resulting line shapes are strongly non-Lorentzian and these effects seem to be accentuated as one proceeds to heavier elements.

7.3 Qualitative Interpretation of ESCA Spectra for N_2, CO, C_2H_2, and C_2H_4

7.3.1 The $1\underline{s}$ Core Hole Spectrum of N_2

The experimental ESCA spectrum by Gelius and coworkers[7, 124)] is shown in Fig. 45 and from the positions of the degenerate HF $1\underline{\sigma}_g$ and $1\underline{\sigma}_u$ levels[120, 125)] one can conclude that the total core-level relaxation shift is ~18 eV. The introduction of monopole relaxation through e.g. a HF $1\underline{\sigma}_{g,u}$ ΔSCF calculation gives a relaxation shift ~7 eV (estimated by comparison with O_2[40)]), leaving about 11 eV to be accounted for by symmetry breaking and localization of the core hole to a $1\underline{s}$-like orbital centred on either nucleus. The entire

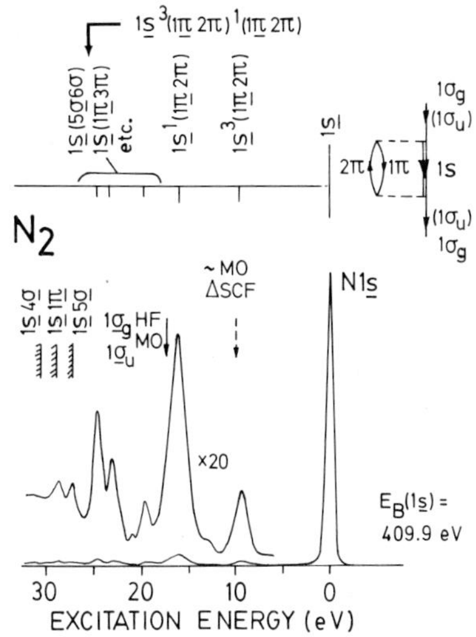

Fig. 45. Experimental core level ESCA spectrum for N_2[7)]

relaxation shift can also be obtained from a ΔSCF calculation with a localized $1\underline{s}$ hole, as demonstrated by Bagus and Schaeffer[40)] in the case of O_2.

In Sect. 7.2.1 we discussed the principles for the g-u symmetry-breaking dipole relaxation of the $1\underline{\sigma}_{g,u}$ levels and the corresponding basic self-energy diagram expressed in terms of the final-state localized orbitals is inserted in Fig. 45. Actually, this self-energy diagram describes the entire relaxation process and the total shift from the HF $1\underline{\sigma}_g$ ($1\underline{\sigma}_u$) position but as long as we select the distorted dipole excitations like $1\pi \rightarrow 2\pi$ ($1\pi_u \rightarrow 1\pi_g$) or $5\sigma \rightarrow 6\sigma$ ($3\sigma_g \rightarrow 3\sigma_u$), the diagram will basically describe the dipole relaxation shift and shake-up satellites. Note that it is very important to have the intermediate $1\underline{s}$ hole *renormalized*, so that relaxation and correlation effects are properly taken into account also in the ionic excited levels.

Turning to the experimental spectrum in Fig. 45, the first two strong satellites at 9.3 eV and 16.0 eV can directly be identified as $1\underline{s}\ ^3(1\underline{\pi}2\pi)$ and $1\underline{s}\ ^1(1\underline{\pi}2\pi)$ respectively. which is very close to the optical excitation energies. The singlet satellite is very intense ($\sim$ 10% of the main line[126)]) and may loosely be referred to as a plasmon shake-up satellite, in analogy with the dominant mode of excitation in an extended electron system. Extending the analogy to the emission of two independent plasmons, the expected excitation energies of the valence π-bond from $\pi \rightarrow \pi^*$ excitations have been marked in Fig. 45, and there is an interestingly good correlation between the predicted and the actual experimental satellite positions. However, it is by no means clear that any significant part of the intensity in the peaks from 20 eV and upwards actually is due to double excitations. There should be fairly important $\sigma \rightarrow \sigma^*$ transitions ($3\sigma_g \rightarrow 3\sigma_u$; $5\sigma \rightarrow 6\sigma$) which might become pushed up in energy by the presence of the $1\underline{s}$ core hole, and there is also a spectrum of Rydberg-like levels (neglecting effects of molecular vibrations) extending below the double-ionization thresholds. From the discussion in Sect. 7.2.1 we may conclude that a reasonable estimate of the screened core-valence Coulomb repulsion is $\sim$ 12 eV, which places the $1\underline{s}5\underline{\sigma}$, $1\underline{s}1\underline{\sigma}$ and $1\underline{s}4\underline{\sigma}$ thresholds in the range 26–30 eV and the $1\underline{s}3\underline{\sigma}$ threshold around 50 eV. This means that there is a high density of ionic excited levels in the region between 20 and 30 eV above the $1\underline{s}$ ionic ground state. This would include excitations like $1\underline{s}\,1\underline{\pi}\,3\pi$[7] and $1\underline{s}\,4\underline{\sigma}\,6\sigma$ which could also be sources of the prominant $1\underline{s}$ satellites in these regions. The CI calculations of Hillier and Kendrick[127] place the $5\sigma \rightarrow 6\sigma$, $4\sigma \rightarrow 7\sigma$ and higher excitations in the 20 to 30 eV range but since the calculated satellite intensities are much lower than experimentally found, the identification of the strong satellites in the 20–30 eV region remains an open question.

Finally, as already pointed out in Sect. 7.2.1, there are a few things to be noted about the satellite intensity distribution and the relaxation shift. With the intensity distribution given by Gelius[7] (corrected for the error in the $\pi \rightarrow \pi^*$ singlet intensity), the discrete part of the spectrum only contributes ~ 5 eV out of a total relaxation shift of ~ 18 eV, demonstrating that a very large part of the relaxation shift is associated with the shake-off satellite spectrum. This means that a substantial part of the continuum background above the shake-off thresholds around 30 eV must be due to a true shake-off continuum and must not be subtracted off when analyzing the experimental spectrum. Furthermore, as seen in Fig. 45 the frozen HF $1\underline{\sigma}_g$ and $1\underline{\sigma}_u$ levels are above the $1\underline{s}\,(\underline{\pi}\pi^*)$ satellites.

7.3.2 The Valence Hole Spectrum of N_2

Assuming that the gross features of the molecular ionic excitation spectrum does not radically change if one substitutes a deep $1\underline{s}$ hole for a valence hole, we can now directly attempt an interpretation of the valence hole spectrum of N_2. Fig. 46 shows the experimental ESCA spectrum of Gelius and coworkers[124] together with the HF MO $1\underline{\pi}_u$, $3\underline{\sigma}_g$, $2\underline{\sigma}_u$ and $2\underline{\sigma}_g$ levels and the uncoupled ionic excited levels based on $\underline{\pi}\pi^*$ singlet-triplet and single-double excitations referred to the exact positions of the $2\underline{\sigma}_u$ and $3\underline{\sigma}_g$ levels (the *experimental* positions of the $1\underline{s}$ satellites referred to the exact $3\underline{\sigma}_g$ and $2\underline{\sigma}_u$ levels,

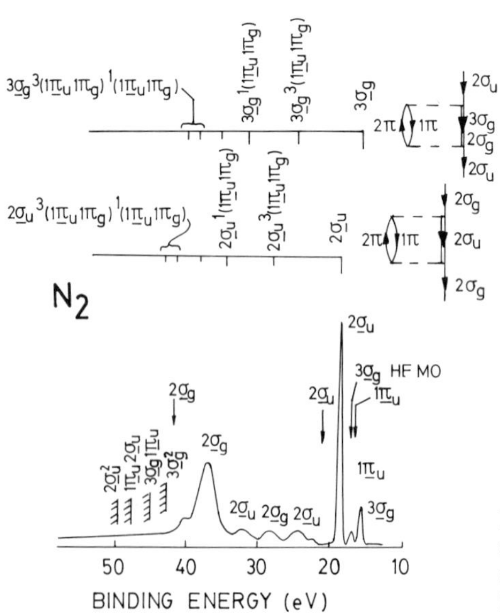

Fig. 46. Experimental valence level ESCA spectrum for N_2[124]. Ionic excitation levels (*top of picture*) inferred from the core hole spectrum in Fig. 45

hopefully representing a realistic estimate of the true ionic level structure, although we do not know at present how to interpret a large part of the structure).

We are now in a position to evaluate the effects of dynamic dipole relaxation, the principles of which have been discussed in Sects. 7.1 and 7.2. The fundamental self-energy diagrams have been inserted in Fig. 46 and the spectrum of levels to the left of the diagram indicates the positions of ionic excited states before being coupled to the primary valence hole. In order to solve for e.g. the $2\underline{\sigma}_u$ spectrum we must know the $3\underline{\sigma}_g$ position which, in turn, requires knowledge of the very $2\underline{\sigma}_u$ spectrum that we are solving for in the first place. Here, we step right into this cycle, *assume* knowledge about one fundamental core level and the ionic excitations built on this level and solve for the spectrum of another level. We then obtain the following characterization of the valence hole spectrum of N_2 (see also[30, 109, 123]):

a) The $1\underline{\pi}_u$ *outer-valence level* essentially has no dynamic relaxation shift and remains delocalized and unshifted. Since the Fermi sea-correlation shift approximately cancels the HF MO ΔSCF shift, the position of the $1\underline{\pi}_u$ level is well represented by the HF MO energy eigenvalue (Koopmans' theorem).

b) The $3\underline{\sigma}_g$ *outer valence level* acquires a dynamic relaxation shift of ~ 2 eV to lower binding energy[30, 123] representing considerable axial distortion (localization) due to mixing with $2\underline{\sigma}_u$. This dynamic localization results in lower binding energy for the $3\underline{\sigma}_g$ level than for the $1\underline{\pi}_u$ level and leads to the well-known reordering of the level structures as compared with both the MO HF (Koopmans' theorem) and the MO HF ΔSCF pictures. The $3\underline{\sigma}_g$ shift is mainly caused by spectral repulsion from the $2\underline{\sigma}_u 1\underline{\pi}_u 1\pi_g$ levels which will give rise to weak satellite structures around 28 eV and from around 35 eV and upwards.

c) The $2\underline{\sigma}_u$ *outer valence level* has a dynamic relaxation shift of ~ 2.5 eV to lower binding energy[30, 123] by fluctuation mainly to $3\underline{\sigma}_g$ and to some extent to $2\underline{\sigma}_g$, leading to appreciable $3\underline{\sigma}_g 1\underline{\pi}_u 1\pi_g$ based satellite lines around 25 eV and 32 eV and to weak structure in the range from 35 eV and upwards. The $2\underline{\sigma}_u$ hole is considerably distorted along the molecular axis by mixing with mainly $3\underline{\sigma}_g$ and this dynamic localization is stabilized by the 1π screening charge. Since the distance between the $3\underline{\sigma}_g$ and $2\underline{\sigma}_u$ levels is much smaller than the typical excitation energies of the π-bond, the screening of the $2\underline{\sigma}_u$ hole (as well as the $3\underline{\sigma}_g$ hole above) is almost static, the screening charge adiabatically following the motion of the hole. Therefore, the quasi-particle picture is perfectly valid for the outer valence holes, and the spectrum is characterized by strong main lines and relatively weak satellite lines.

d) The $2\underline{\sigma}_g$ *inner valence level* behaves very differently from the outer valence levels because the frequency of fluctuation of the $2\underline{\sigma}_g$ hole is now approximately equal to the strongest dipole resonances of the π-bond. The $2\underline{\sigma}_g$ hole and the 1π-screening charge then form a strongly coupled system and the spectrum becomes broken up into a number of more or less complicated normal modes. From Fig. 46 we see that the $2\underline{\sigma}_g$ level is nearly degenerate with a number of $2\underline{\sigma}_u$-based ionic excited levels and, qualitatively, it is easy to realize that the $2\underline{\sigma}_g$ level will become highly distorted and spread over a number of ionic levels. From a quantitative point of view, however, the problem is not fully understood at present. The calculation by Schirmer et al.[109] clearly demonstrates that the coupling strength is large enough to strongly distort and broaden the $2\underline{\sigma}_g$ peak and make plausible that it consists of several nearly equally strong peaks, giving the experimental peak its flattened and asymmetric shape. However, the actual shape of the main $2\underline{\sigma}_g$ peak, as well as its position and the satellite structure in the range 30–45 eV are not correctly described by Schirmer et al.[109], suggesting that while the average density of ionic excited states may be reasonably well described, the individual levels are not. One example of this is that Schirmer et al.[109] seem to place the experimental 32 eV $2\underline{\sigma}_u$ satellite at much too high energy up in the low energy flank of the $2\underline{\sigma}_g$ complex.

Judging from experiment (Fig. 46), the $2\underline{\sigma}_g$ total relaxation shift is ~ 5 eV (assuming a Fermi-sea correlation shift of ~ -1 eV), of which Schirmer et al.[109] account for ~ 2.5 eV, placing the main $2\underline{\sigma}_g$ level structure at ~ -40 eV. In the case of the $1\underline{s}$ core level we noticed that most of the contribution to the relaxation shift came from the continuum of excited states and it is just possible that the $2\underline{\sigma}_g$ hole is deep enough that a limited basis set in the self-energy calculation will make it very difficult to reproduce the full shift. The situation probably corresponds to that illustrated in Fig. 44 a (case (3)) and in Fig. 44 d, the $2\underline{\sigma}_g$ level becoming shifted by high-lying discrete and continuum excitations down to where it becomes degenerate with the $2\underline{\sigma}_u{}^1(\underline{\pi}\underline{\pi}^*)$ ionic excited level as well as with a number of other levels.

Finally, it should be mentioned that the present assignment of the peak structure in Fig. 46 agrees with the recent measurements by Weigold et al.[128]. Therefore, the gross features of the valence hole spectrum of N_2 can be said to be fairly well understood even though the region above 20 eV cannot yet be well reproduced by first principles calculations.

7.3.3 The 2 σ (C1s) Core Hole Spectrum of CO

The CO molecule has the same number of electrons as N_2 but because it is heteronuclear, the bonds are markedly localized towards either nucleus. All holes are therefore created predominantly on one nucleus or the other, and a HF MO ΔSCF calculation will directly give a larger fraction of the total relaxation energy than in N_2. In particular, the HF MO scheme gives the correct ordering of the valence hole levels in contrast to the case of N_2.

Having discussed hole level spectra in N_2 in considerable detail, we only give a brief presentation of the corresponding results in CO and a few other systems. Fig. 47 shows

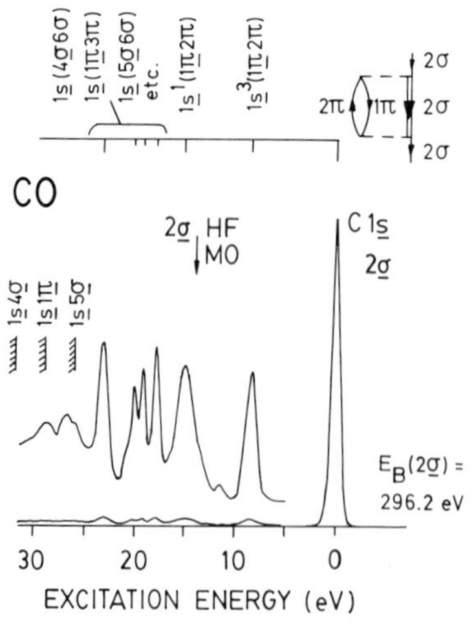

Fig. 47. Experimental core level ESCA spectrum for CO[7]

the experimental CO $2\underline{\sigma}$(C$\underline{\text{l}}$s) ESCA spectrum by Gelius and coworkers[7]; from the HF MO $2\underline{\sigma}$ position[120, 125] one can deduce a total relaxation shift of ~15 eV. Just as for N_2 there are two dominating low-energy satellites which we identify as $2\underline{\sigma}^3(1\underline{\pi}2\pi)$ at ~8 eV and $2\underline{\sigma}^1(1\underline{\pi}2\pi)$ at ~15 eV, as shown in Fig. 47. The tentative mechanism of simultaneous shake-up of two independent $\pi \to \pi^*$ excitations will again coincide with a very strong high-energy satellite line. However, it could also (and perhaps more likely) be that this line is a single excitation to the next higher π-level, $1\pi \to 3\pi$, as suggested by Gelius[7], or is the first strong line in a series converging to e.g. the $2\underline{\sigma}4\underline{\sigma}$ threshold, like $4\sigma \to 6\sigma$.

The lower axial symmetry and the localization of orbitals in the CO molecule is clearly reflected in the intensity distribution of the $2\underline{\sigma}$ satellite spectrum. In comparison with N_2, the low-lying $2\underline{\sigma}(1\underline{\pi}2\pi\,^1\Sigma)$ charge-transfer type of satellite loses much of its intensity and instead other more high-lying levels, probably corresponding to shake-up transitions predominantly localized to either nucleus (e.g. $1\pi \to 3\pi$, $4\sigma \to 6\sigma$, $5\sigma \to 7\sigma$) grow in importance. Also, the relative intensity of the discrete part of the satellite spectrum with respect to the main line seems to be weaker in CO than in N_2, indicating that more strength has gone into continuum excitations, like in atomic shake-up and shake-off spectra.

7.3.4 The Valence Hole Spectrum of CO

Again proceeding like in the case of N_2, the experimental CO valence ESCA spectrum by Gelius et al.[124] is shown in Fig. 48 together with an estimate of the $\pi \to \pi^*$ excitation spectrum and the actual experimental $2\underline{\sigma}$ satellite spectrum referred to the $5\underline{\sigma}$ and $6\underline{\sigma}$ levels. Because the occupied orbitals are predominantly localized on either nucleus, the MO SCF scheme[120, 125] gives directly the right ordering of the valence hole levels and the ΔSCF procedure leads to binding energies in reasonable agreement with experiment.

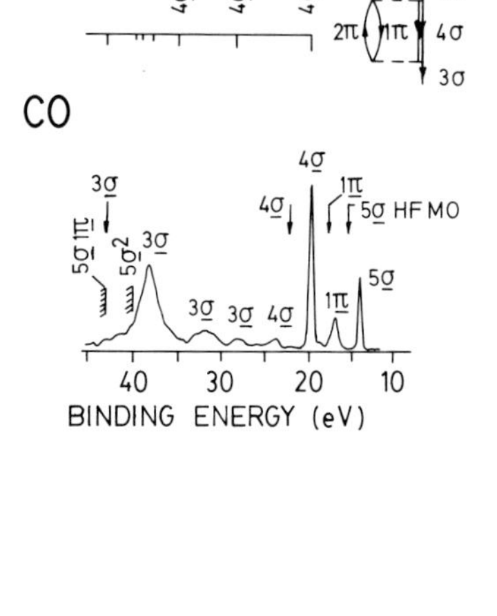

Fig. 48. Experimental valence level ESCA spectrum for CO[124]. Ionic excitation levels (*top of picture*) inferred from the core hole spectrum in Fig. 47

The ΔSCF static relaxation shift for the $5\underline{\sigma}$ level was obtained from Bagus and Viinikka[112], $\Delta_{5\underline{\sigma}}^{S0} \simeq 1.8$ eV, and we have assumed it to be valid for the other outer valence hole levels as well. The $3\underline{\sigma}$ ΔSCF level shift has been estimated from inspection of the results of Schirmer et al.[109] to be ~2.5–3.0 eV.

The fact that the ΔSCF positions of the *outer valence hole* levels agree rather well with experiment shows immediately that the dynamic relaxation process (valence hole fluctuation) is far from negligible, being of the order of the Fermi-sea correlation shift and opposite in sign, i.e. roughly 1–2 eV. The dynamic correlation between a localized hole and the valence screening charge is therefore still very important, although the bonds themselves are more ionic in character, and there is no dramatic change in comparison with the case of N_2. For the *inner valence* $3\underline{\sigma}$ hole the situation is less clear since there is no accurate MO ΔSCF binding energy to compare with. However, the more line-like appearence immediately suggests that the coupling to the ionic excitation spectrum is considerably weaker than in N_2 and that a relatively larger part of the relaxation shift is made up of static relaxation, as can be directly obtained from a MO ΔSCF calculation.

Regarding the satellite structure, the main features look very much the same as in N_2 with pronounced peaks at around 23.5 eV, 28 eV and 32 eV, suggesting that the ionic excitations are similar in N_2 and CO. Also in CO all of the major satellites in the valence region correspond to a major feature in the $2\underline{\sigma}$ core hole satellite spectrum and by finding out which core hole is mainly driving a particular ionic excitation, e.g. through the self-energy diagrams in Figs. 48 c, d, one should be able to assign any particular satellite peak to one or several parent valence holes. The results of Bagus and Viinikka[112] are then very interesting, ascribing most of the satellite intensity in the region from 25 to 45 eV to the $3\underline{\sigma}$ inner valence level. Moreover, most of the satellite intensity is predicted to go into *double excitations,* as indicated in Fig. 48.

It is very easy to believe that most of the satellites derive their strength from the $3\underline{\sigma}$ level because, due to the low axial symmetry, this level will couple to all kinds of excitations, just like a deep core level. Moreover, it is more than likely that double excitations play an important role in the description of some of the satellites and the structure of the main $3\underline{\sigma}$ peak. However, for a more precise characterization of the valence spectrum, the quality of the existing calculations[109, 112] does not seem to be sufficient. As clearly stated in[112], the restriction to a model space of valence orbitals seriously underestimates the static relaxation shift (cf. Sect. 7.1) which may change the relative positions of the levels and consequently also the relative intensities and assignments. Also in[109] the $3\underline{\sigma}$ peak is placed at too high energy and the spectrum is constructed exclusively by coupling the single valence hole levels to *ionic single excitations.* Both Bagus and Viinikka[112] and Schirmer et al.[109] have obtained a reasonable description of the CO valence spectrum and it would be very interesting to know how to reconciliate these, as it seems, disparate points of view.

7.3.5 The Core and Valence Level Spectrum of Acetylene

Acetylene, C_2H_2, is isoelectronic with N_2 and is the simplest triple-bonded hydrocarbon. It has the same ordering of the HF MO hole levels as N_2, as shown in Fig. 49, but there are some important differences because the nuclear charge is more spread out in C_2H_2 and the potential on the carbon sites is less attractive. Therefore, the $1\underline{\pi}_u$ level in C_2H_2

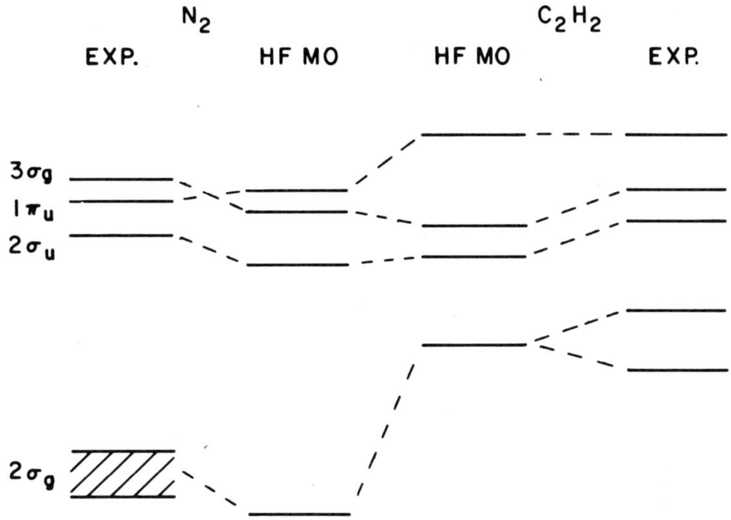

Fig. 49. Comparison of HF MO and experimental valence energy levels for the isoelectronic molecules N_2 and C_2H_2

appears at considerably less binding energy than the other outer valence levels and the $2\underline{\sigma}_g$ level has moved down from ~42 eV in N_2 to ~28 eV in C_2H_2. However, the $2\underline{\sigma}_u$ and $3\underline{\sigma}_g$ orbitals, which are engaged in bonding the hydrogens, remain essentially unshifted in comparison with N_2. The $1\underline{\pi}_u$ HF level lies so much higher than the $3\underline{\sigma}_g$ level that dynamic relaxation effects can do nothing to change this picture. Therefore, in contrast to the case of N_2, in C_2H_2 the HF MO picture gives the correct ordering of the valence levels.

Regarding the character and magnitude of the relaxation and correlation effects, the core region and the outer valence regions in C_2H_2 (Figs. 50, 51) are very similar to those of N_2 (Sects. 7.3.1 and 7.3.2) and the shifts of the corresponding levels are nearly the same. For the inner valence $2\underline{\sigma}_g$ level, however, the picture is quite different because in C_2H_2 this level has moved down into the region of the $2\underline{\sigma}_u\,^3(\underline{\pi}\underline{\pi}^*)$ ionic excitation, and the spectrum looks rather simple with a main line and a very prominent satellite line at higher binding energy.

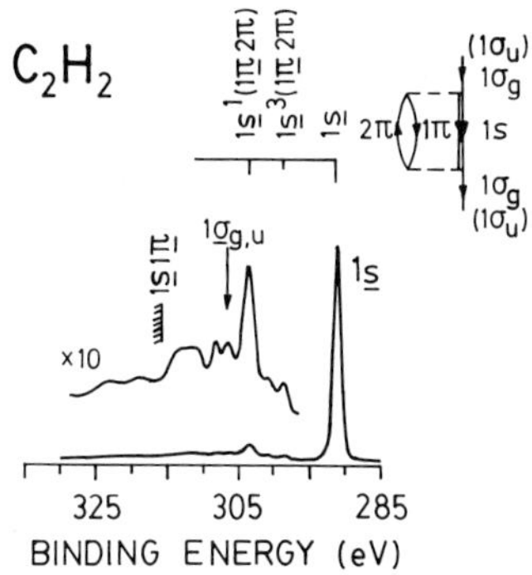

Fig. 50. Experimental $1\underline{s}$ core level XPS spectrum for C_2H_2[129], Mg K_α(1253.6 eV)

The recent experimental core level spectrum by Cavell and Allison[129] in Fig. 50 is very similar to that of N_2 (Fig. 45), with a weak $1\underline{s}\ ^3(1\underline{\pi}^2\pi)$ satellite with $\sim 2\%$ relative intensity at 7.2 eV above the main line, a strong $1\underline{s}\ ^1(1\underline{\pi}2\pi)$ satellite with $\sim 10\%$ relative intensity at 12.2 eV, followed by a prominent discrete and continuous satellite spectrum at higher energies. The total relaxation shift is ~ 16 eV, the MO HF $1\underline{s}$ level lies well above the low-lying $\underline{\pi}\pi^*$ satellites and the limited range of the shake-up and shake-off spectrum shown in Fig. 50 only accounts for about one third of the relaxation shift (through Eq. (22)).

The $\underline{\pi}\pi^*$ triplet and singlet excitations at 7.2 eV and 12.2 eV (to be compared with the optical excitation energies at ~ 5 eV and ~ 9 eV) should also represent reasonable estimates of the $\underline{\pi}\pi^*$ excitation energies relative to any other core level, e.g. the $2\underline{\sigma}_u$ level as shown in Fig. 51. The interpretation of the $2\underline{\sigma}_g$ region of the valence spectrum then becomes straightforward. Due to static and dynamic relaxation and Fermi sea correlations, excluding the $2\underline{\sigma}_u\ ^3(1\underline{\pi}_u 1\pi_g)$ ionic excited state, the $2\underline{\sigma}_g$ hole level will be shifted down to around 25–26 eV binding energy. There it becomes nearly degenerate with the $2\underline{\sigma}_u\ ^3(1\underline{\pi}_u 1\pi_g)$ ionic excited level and is split into two prominent lines at around 23.5 eV and 27.5 eV with relative intensities of about 2:1. There is a fairly delicate balance involved because raising the $2\underline{\sigma}_g$ HF MO level by a few eV will reverse the intensity ratio of the two lines. The question which line represents the $2\underline{\sigma}_g$ level is therefore not particularly meaningful and the spectral shape of this region could vary considerably between different systems although the coupling mechanism and basic physics would be the same (for further discussion and examples see[110]).

Dealing with a case of nearly degenerate $2\underline{\sigma}_g$ and $2\underline{\sigma}_u\ ^3(1\underline{\pi}_u 1\pi_g)$ configurations one may ask what kind of spacial correlations are involved. In Sects. 7.1 and 7.2 we discussed the breaking of the gerade-ungerade symmetry of the molecule and dynamic localization of the hole, i.e. hopping of a localized hole between the nuclei. That discussion was based

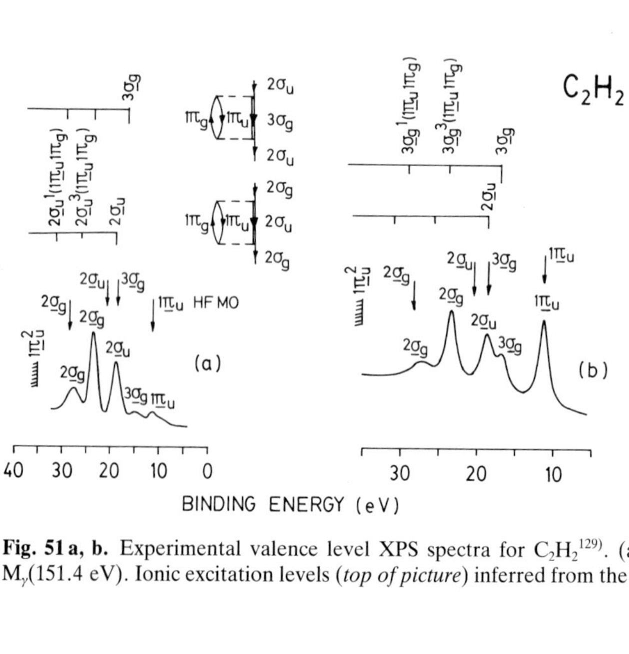

Fig. 51 a, b. Experimental valence level XPS spectra for C_2H_2[129]. (a) Mg K_α(1253.6 eV), **(b)** Zr M_γ(151.4 eV). Ionic excitation levels (*top of picture*) inferred from the core hole spectrum in Fig. 50

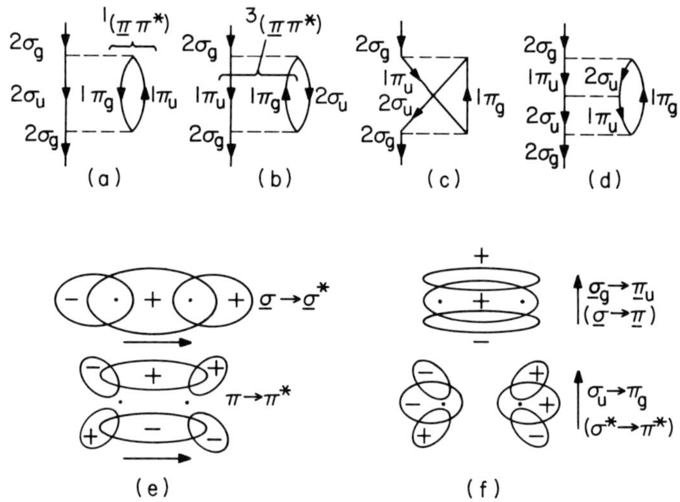

Fig. 52. (a)–(d) Self-energy for an inner valence $2\underline{\sigma}_g$ hole, describing the excitation of triplet and singlet $\underline{\pi}\underline{\pi}^*$ satellites. **(e)** and **(f)** suggest that the singlet $\underline{\pi}\underline{\pi}^*$ excitation is associated with distortions and dipole moments along the molecular axis **(e),** and the triplet $\underline{\pi}\underline{\pi}^*$ excitation with distortions and dipole moments perpendicular to the axis **(f)**

on the self-energy diagram in Fig. 52a involving screening of a $2\underline{s}$-like hole by an axial displacement of π-charge, i.e. involving singlet $\underline{\pi}\underline{\pi}^*$ excitations. This process does not admit excitation of the triplet except via the spin-orbit interaction, which, however, is far too weak to play any role here (cf. optical spectra). However, a spin-flip can be induced by the Coulomb interaction through an exchange process, as discussed e.g. in Sect. 6.1.1. In this way, Fig. 52b describes the direct excitation of the $2\underline{\sigma}_u\,{}^3(1\underline{\pi}_u 1\pi_g)$ level and since all orbitals involved belong to the valence space, the excitation matrix element should be large.

Concerning the spatial picture, the important point about Fig. 52b is that the $2\underline{\sigma}_g$ hole now fluctuates to $1\underline{\pi}_u$, *not* to $2\underline{\sigma}_u$ and, as a consequence, the charge displacement is *perpendicular* to the molecular axis (Fig. 52f) instead of parallel to the axis (Fig. 52e). This implies also that the rotational symmetry with respect to the molecular axis is broken for the one-electron states and only valid for the total many-body state. The perpendicular dipole-dipole correlation is precisely analogous to the atomic case, as discussed in Sect. 6. The hole has now freedom to distort its orbital and to become more or less localized also perpendicular to the molecular axis. The frequency of this motion is given by the energy difference between the $1\underline{\pi}_u$ and $2\underline{\sigma}_g$ levels and is evidently much higher than the frequency of motion parallel to the axis. However, since the singlet and triplet $\underline{\pi}\underline{\pi}^*$ excitations represent different internal couplings of the same basic configuration $2\underline{\sigma}_u(\underline{\pi}\underline{\pi}^*)$, they can interfere and mix, through e.g. the exchange process in Fig. 52c and the hole-hole exchange scattering process in Fig. 52d. This will mix the parallel and perpendicular modes and shift the singlet and triplet excitation energies. Expressed differently, proper coupling in of an additional *valence* hole can shift and split the singlet and triplet $(\underline{\pi}\underline{\pi}^*)$ excitations and transfer intensity between them.

7.3.6 The Core and Valence Hole Spectra of Ethylene

Ethylene, C_2H_4, is the simplest *double-bonded* hydrocarbon and has the same number of electrons as O_2. The energy level diagram in Fig. 53 schematically shows the effect of lowering the rotational symmetry of O_2 and lifting the degeneracy of the π-electrons, leading to different energies for the in-plane (π_y, $\pi_\parallel$) and the out-of-plane (π_x, $\pi_\perp$) π-orbitals. Proceeding to C_2H_4, the $1\pi_u$ and $1\pi_g$ orbitals both become strongly hybridized and develop into σ-like $1b_{2u}$ and $1b_{3g}$ C_2H_4 symmetry orbitals (Fig. 53), the π-charge becoming transferred from the C–C regions. Furthermore, the $1\pi_g$-derived orbital is now completely filled and there are then no longer any low-lying $\pi \to \pi^*$-like transitions involving the in-plane valence charge. The out-of-plane π-orbials, on the other hand, give rise to a low energy $\pi_\perp \to \pi_\perp^*$ transition. However, since the $\pi_\parallel \to \pi_\parallel^*$ transitions no longer exist, the co-operative dipole-dipole interaction among the π-electron is weakened and the collective shift (singlet-triplet splitting) becomes reduced in comparison with the doubly π-bonded acetylene molecule. Moreover, since now only half as many π-electrons take part in the $\pi \to \pi^*$ transitions, the spectral weight of such satellites should become reduced in comparison with doubly π-bonded molecules like N_2 and C_2H_2.

Figure 54 shows the $1\underline{s}$ XPS spectrum of C_2H_4 by Carlson et al.[130)]. The prominent bump around 20 eV is found also in the saturated, π-bonded hydrocarbons but the peak at 8.4 eV is specific to the unsaturated hydrocarbons and has been associated with the out-of-plane $\pi \to \pi^*$ transitions[130)]. In view of its large intensity (9.7% of the main peak), we associate this peak with the singlet ionic excitation $1\underline{s}^1(1\underline{b}_{3u}\,1b_{2g})$ (disregarding the broken symmetry; in simplified notation $1\underline{s}^1(\underline{\pi}\pi^*)$), in direct analogy with the $1\underline{s}$ spectra of C_2H_2 and N_2. The triplet excitation could be expected to lie around 5 eV and could possibly have something to do with the shoulder at the base of the main $1\underline{s}$ line. The very prominent satellite bump at ~20 eV must have a very composite character involving a multitude of excitations and probably corresponding to an average excitation of the in-plane σ-like charge distribution (into which the $\pi_\parallel$ and $\pi_\parallel^*$ charge density was redistributed), while the 8.4 eV satellite represents the major excitation of the out-of-plane π-like charge distribution.

It is very interesting to compare this with the $1\underline{s}$ core level spectrum of O_2[131)], which can be regarded as the equivalent (united) diatomic molecule of C_2H_4. Much in the same

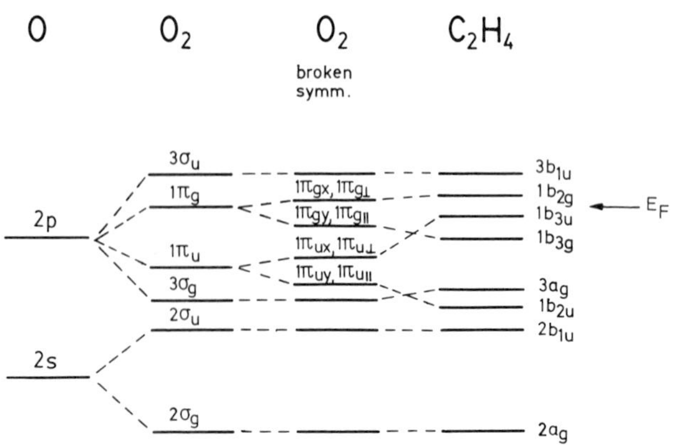

Fig. 53. Schematic picture of the development of the one-electron energy levels of C_2H_4 from breaking of the spherical symmetry of the isoelectronic O_2 molecule

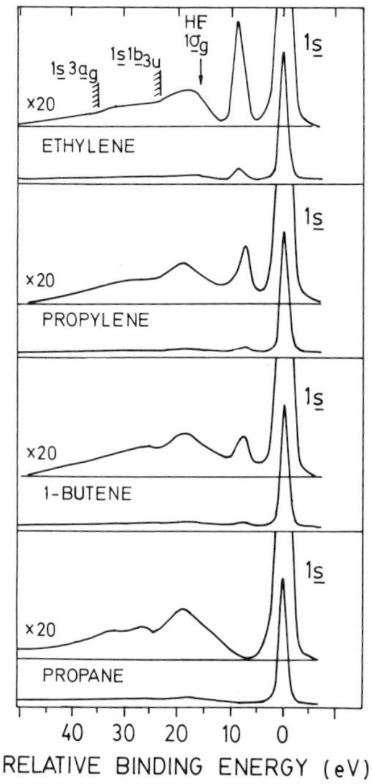

Fig. 54. Experimental core level XPS spectra of C_2H_4, C_3H_6, C_4H_8 and C_3H_8[130])

way as the C_2H_4 spectrum in Fig. 54, the O_2 satellite spectrum is dominated by a prominent $\pi \rightarrow \pi^*$ excitation at around 10 eV and another prominent excitation around 20 eV and above. It seems that in O_2 the $^1(\underline{\pi}\pi^*)$ satellite has a higher excitation energy and possibly a higher relative intensity than in C_2H_4, which may reflect stronger collective effects, and it also appears to be stronger relative to the 20 eV peak, pointing towards a less important role of the σ-like charge distribution in O_2 than in C_2H_4 and the higher alkenes.

Figure 55 shows the experimental valence spectrum of C_2H_4[7]) together with the positions of the MO HF levels and the estimated positions of the double ionization thresholds involving at least one $1\underline{b}_{3u}(\underline{\pi}_{\perp})$ hole. At XPS photon energies, the outer valence 2p-derived levels have much lower photoionization cross sections than the inner and outer valence 2s-derived levels $2\underline{a}_g$ and $2\underline{b}_{1u}$ and therefore appear very weak. Just as in the case of C_2H_2, the outermost π-level ($1\underline{b}_{3u}$, $\underline{\pi}_{\perp}$) is shifted to higher binding energy which demonstrates that the Fermi sea ground-state correlations dominate for this level. For the deeper outer valence levels, as a rule of thumb, we consider the ΔSCF shift and the Fermi sea correlation shifts approximately to cancel and we take the deviations from the HF MO values to indicate the size of the shifts due to dynamic relaxation (fluctuation, localization).

The $1\underline{b}_{3u}$ ($\underline{\pi}_{\perp}$) is the *only* valence hole that cannot fluctuate, there being no occupied π levels for the hole to fluctuate to. It therefore remains delocalized, and since the monopole relaxation shift is very small, the level shift is dominated by Fermi sea correlations. All the other valence orbitals can fluctuate and have substantial relaxation shifts due to

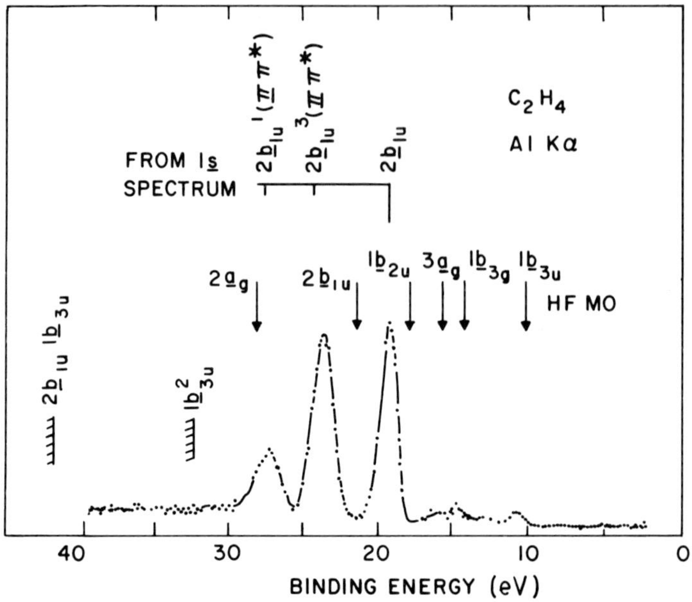

Fig. 55. Experimental valence level ESCA spectrum of C_2H_4[7,8]

localization. Fig. 56 shows some fundamental fluctuation modes in the form of the corresponding lowest order self-energy diagrams.

Compared with experiment[7] (Fig. 55) it seems that the $1\underline{b}_{3g}$, $3\underline{a}_g$ and $1\underline{b}_{2u}$ outer valence holes become shifted by 1–1.5 eV and the $2\underline{b}_{1u}$ $(2\underline{\sigma}^*)$ inner valence hole by more than 2 eV due to fluctuation/localization, precisely along the lines previously discussed for N_2 and C_2H_2. Concerning the $2a_g$ inner valence region, it looks very similar to the $2\underline{\sigma}_g$

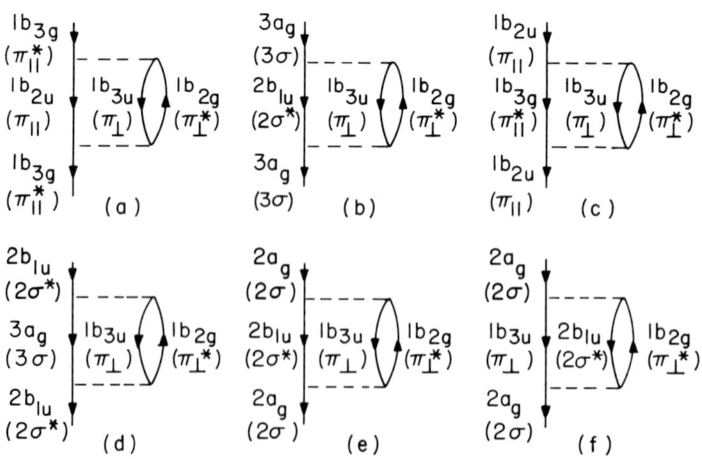

Fig. 56 a–f. Self-energy diagrams describing the most important satellite excitations in the valence region of C_2H_4 (for further explanations, see text; also cf. Fig. 40)

region in C_2H_2, with a main line showing a total relaxation shift of more than 5 eV and a prominent satellite line with about 40% of the intensity of the main line. The explanation is then most likely the same, namely fluctuation of the $2\,\underline{a}_g$ hole to $2\underline{b}_{1u}$ ($\sigma \rightarrow \sigma^*$) coupled to fluctuation of the $1\,b_{3u}$ electrons to $1\,b_{2g}$ ($\pi \rightarrow \pi^*$) as shown in Fig. 56e. Due to the near-degeneracy of the $2\,\underline{a}_g$ and $2\underline{b}_{1u}$ $^3(1\underline{b}_{3u}\,1\,b_{2g})$ levels, the coupling becomes very strong and the resulting levels obtain comparable intensities (Fig. 55). Furthermore, there will be coupling to the singlet $\underline{\pi}\pi^*$ excitation, giving rise to a weak satellite around 31 eV. This explanation was first given by Martin and Davidson[42)] and is basically supported by the more detailed calculation by Cederbaum et al.[111)] which however suggests that the satellites consist of several lines.

In the present case of C_2H_4 the $1\,\underline{s}$-core level spectrum turns out not to be very helpful when discussing the $2a_g$ inner valence spectrum, because first of all there is no clear indication to the whereabouts of the $^3(\underline{\pi}\pi^*)$ excitation. Secondly, the singlet excitation and the estimated triplet excitation seem to have too low excitation energies (see Fig. 55), being only a little larger than the optical triplet and singlet excitation energies of 4.6 and 7.6 eV. The results of[42, 111)] rather point towards 7 and 10 eV for the triplet and singlet $\underline{\pi}\pi^*$ excitations in the presence of a $2\underline{b}_{1u}$ hole. This may reflect the importance of a proper treatment of the ionic excitation, taking into account the difference between a deep hole and a valence hole regarding the average Coulomb interaction as well as angular momentum coupling.

7.3.7 *Validity of the Molecular Orbital and Quasi-Particle Concepts*

Let us now summarize some of the results for the real systems investigated so far in terms of the molecular orbital and quasi-particle properties of the core and valence levels. Starting from a HF MO picture, one can have the following sequence of characteristics:
a) Validity of the MO picture implying validity of the quasi-particle concept.
b) Breakdown of the MO picture *without* breakdown of the quasi-particle picture. Implies lack of symmetry adaption of the hole.
c) Partial breakdown of the quasi-particle picture in the sense that the hole level becomes distributed over several strong components. Indicates breakdown of the MO picture.
d) Complete breakdown of the quasi-particle picture in the sense that there are no longer any well-defined discrete features that can be associated with any kind of elementary excitation.

These characteristics can now be associated with the core and valence hole levels in the following way:
a) The MO picture only applies to those very few outer valence levels that cannot fluctuate to occupied levels with similar spatial extent but with different inversion (g–u) symmetry. Note that it does not have to be the outermost level ($1\,\underline{\pi}_u$ in N_2).
b) In general, an MO picture is not valid because localization effects cause relaxation shifts that are of the order of or larger than those induced by delocalized holes. For a core level, the result is extreme *static* localization to either central core, while for the outer valence levels the localization is *dynamic* in the form of an oscillating disturbance (in a time picture) of the charge density distribution of an MO-hole. Exactly, how drastic this

localization is remains unknown at present, but since it contributes the largest part of the relaxation shift it has to be significant. Whether the shift can be obtained through a ΔSCF calculation with a localized hole remains to be seen. In principle, it could be handled within the unrestricted-HF scheme.

The above-mentioned breakdown of the *HF MO picture* does not imply that the *quasi-particle* picture has broken down, because the outer valence and the core spectra are dominated by a strong main line although the satellite intensity can be appreciable. Therefore, there exists a well-defined ionic ground-state molecular orbital although it might be more or less distorted and localized in comparison with the neutral molecule.
c) This is the situation relevant to the $2\underline{\sigma}_g$ ($2\underline{a}_g$) inner valence levels. There seems to be a general non-resonant radial and angular (symmetry breaking) relaxation and on top of that very strong coupling between a number of nearly degenerate levels. In the hydrocarbon series this gives rise to what seems to be two specific modes where a distortion of the $2\underline{\sigma}_g$ hole perpendicular to the molecular axis oscillates in phase and out of phase with a distortion (screening charge) of the $2\underline{\sigma}_u$ orbital perpendicular to the axis. The in-phase mode becomes excited with the largest intensity and gives rise to the main line while the out-of-phase mode shows up as a very strong satellite at higher binding energy. Although this represents complete breakdown of any conventional MO picture, the spectral peaks do correspond to fairly well-defined elementary excitations of the hole level and do not represent any breakdown of the quasi-particle picture, unless defined in a very narrow sense. Instead, we are observing and dealing with the normal modes of two strongly coupled oscillators immersed in a polarizable medium.

In principle, the $2\underline{\sigma}_g$ region in N_2 could be equally simple, basically reflecting the normal mode spectrum of two strongly coupled oscillators consisting of a $2\underline{s}$-like hole oscillating between the nuclei in-phase and out-of-phase with its screening cloud of distorted π-electron charge. In practice, however, there is a high density of other excitations in this region and the experimental $2\underline{\sigma}_g$ peak in Fig. 46 gives rather the impression of being located in a quasi-continuum of prominent levels that will give strong dispersion and broadening effects, much in the same way as discussed in Sect. 6.1.2. Whether one should call this breakdown of the *quasi-particle* picture, and not only breakdown of the MO picture, then becomes largely a matter of taste, the complicated structure only showing up in the long-time behaviour. However, there are a number of molecular systems exhibiting complete breakdown of the quasi-particle picture in the inner valence region, as demonstrated by Cederbaum et al.[117, 118].

8 Quasi-Particle Properties of Hole Levels in Solids and Adsorbate Systems

8.1 Overview

This section will contain less analytical and more straightforward presentation of current experimental and theoretical results and ideas. The field is developing extremely rapidly and fairly little has been done so far in terms of quantitative calculations on real systems. Most of the theoretical investigations have been in the form of one-electron calculations for molecules and clusters, or many-body calculations on model systems. We shall discuss a number of cases where hole spectra in extended systems are strongly influenced or, in certain regions, even dominated by atomic-like, local excitations, with the particular aim of presenting an overview of unifying aspects.

One interesting case is associated with the response of a system to the creation of a localized hole which breaks the translational symmetry, and how the spatial structure and the excitation spectrum of the localized screening charge differ between e.g. a free-electron metal, a 3 d-transition metal or metal complex, and a 4 f-rare earth metal or metal complex. Another very important case is connected with dynamic localization of valence holes in narrow-band transition metals, due to the screening response of the system. It seems likely that a 3 d-hole in Ni metal is strongly affected by intra-3 d-band fluctuation and decay processes, leading to considerable localization and band narrowing and to well-defined shake-up satelite structures. A third case concerns screening and relaxation of localized holes in atoms and molecules adsorbed on metal surfaces, and the role of surface plasmon versus charge transfer screening for determining photoelectron spectral intensities.

Characteristic of these cases is that the Coulomb potential of the hole may be so strong as to pull empty atomic-like levels down below the Fermi level (cf. the case of atomic Ba, Sect. 6.4). This introduces a strong degeneracy, and the electronic structure of the N-electron ground state might no longer be relevant to the description of the ground and excited states of the ionic N-1 electron system in the neighbourhood of the hole. A typical case is charge-transfer screening in a transition metal complex or adsorbate system where in the ionic ground state screening charge has been transferred from the ligands to the central metal ion, or from the metal surface to the adsorbed molecular ion.

An understanding of relaxation phenomena and response to charge impurities or deep holes seems to have developed independently in several fields like atomic physics, chemistry, solid-state and surface physics. The fundamental problem of screening a charged impurity in free-electron-like metals[23, 27, 28] was discussed long ago by Friedel[132]. More recently, interest has focussed on shake-up effects, threshold singularities and core line asymmetries[47, 133–138]. This field has expanded enormously both theoretically and experimentally, and has seen many heated debates. For the later and most recent development the reader is referred to review articles[48, 49, 139, 140] and to some very recent papers[141–144] and references therein. Here, we shall instead discuss localization of screening charge into atomic-like orbitals in metals and compounds involving transition metals (TM), rare-earth metals (REM) and adsorbate systems. We shall use models having the common feature of empty, atomic-like levels being pulled down by the core hole potential to the vicinity of or below the Fermi level ([14, 24, 47, 145–148] and references cited therein). When occupied, these localized orbitals or resonances form very compact atomic or molecular-like screening clouds and the associated relaxation shifts are larger or even much larger than in the case of screening by free electrons[149].

Figure 57 gives an overview of some interesting cases. In atomic Ba (Fig. 57 a), a diffuse 6 s-shell can collapse into a more compact 5 d-shell; in TM, REM and 5 f-group charge-transfer complexes (Fig. 57 d) screening charge can collapse from a diffuse shell centred on the ligands to a compact 3 d, 4 f or 5 f-shell centred on the metal ion[20, 65, 150–164]; in adsorbate systems (Fig. 57 c) screening charge can be transferred from the metal surface to a compact molecular orbital localized on the adsorbate[50, 91, 147, 165–171]. Screening in the REM's might also be characterized as charge transfer because the 4 f-electrons are non-conducting and localized so that the screening charge has to jump into the 4 f-shell from more diffuse metallic valence levels (Fig. 57 b). In atomic Ba, the REM's and the REM charge-transfer complexes, the coupling between the localized level and the diffuse or extended levels is relatively weak so that even if the localized level is pulled down into the Fermi sea below the Fermi level, there is a large probability that it will remain empty. E.g., in the La 3 d-XPS metal spectrum[10, 172, 173] the lowest level of the system with a 3 d-hole, corresponding to the 4 f-level below the Fermi level being filled, appears as a weak shake-down satellite structure on the low-binding

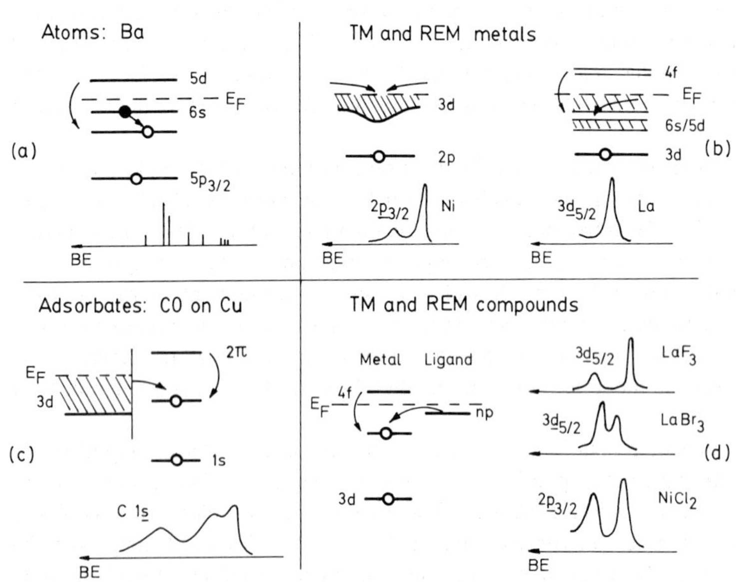

Fig. 57 a–d. Overview of cases where the creation of a core hole causes important reorganization of the electronic valence level structure (for further explanations see text)

energy side of the most intense spectral line. Whether the strongest line should also be called the main line even when it corresponds to an excited state of the hole level can by subjected to debate. There are cases where the distinction between main line and shake-up or shake-down satellites becomes entirely meaningless if intensity is used as a criterion. In LaF_3 we may talk about a 3 $\underline{d}$ shake-up satellite but in $LaCl_3$, $LaBr_3$ and La_2O_3 (e.g.[159, 160]) as well as in 3 d-TM complexes like CoO (e.g.[152]), the "main" line and the "satellite" have about equal intensities. It may then be a better operational definition to associate the main line with the lowest level of the hole, regardless of its spectral intensity.

It seems that cases of strong coupling can be found in e.g. 3 d-TM's and probably also in 3 d-TM compounds and in many adsorbate systems. In such cases, the overlap between the "localized" orbital and the valence orbitals is quite large. The screening process then rather corresponds to a pronounced shift of the bonding charge towards the central metal ion or towards the adsorbate, and the lowest line picks up most spectral strength and also becomes the main line.

Static screening of deep holes in extended systems can be discussed in terms of one-electron ΔSCF type of calculations by breaking the translational symmetry and localizing the hole to an atomic orbital on a particular atom. However, when the hole is created in a narrow band, a localized hole wave-packet will move and the screening will be dynamic, the screening cloud following the motion of the hole and forming a quasi-hole. If the quasi-hole wave-packet moves fast enough it can decay by dissipating energy through excitation of real electron-hole pairs. Dynamic screening and decay of a moving hole cannot be described within a one-electron picture and is intermediate between a symmetry breaking localized hole and an extended hole with the symmetry of the infinite solid. Dynamic screening can be expressed in terms of fluctuation (configuration interaction) and decay processes; the damping of the quasi-hole from the emission of electron-hole pairs corresponds to intra-band type of Auger processes, i.e. Coster-Kronig and giant Coster-Kronig decay processes.

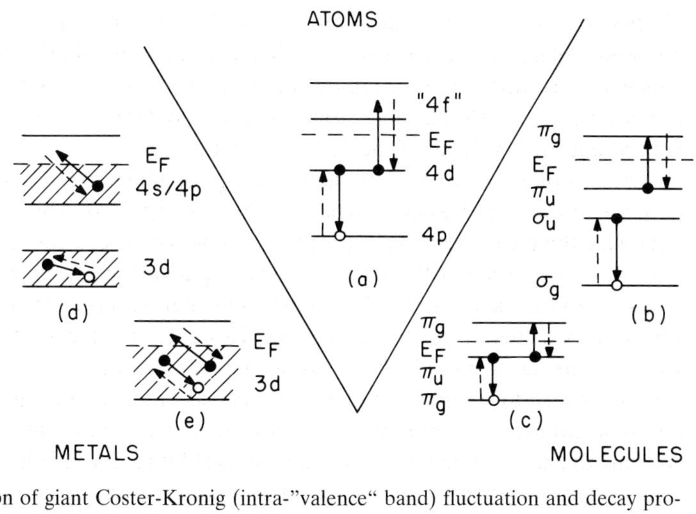

Fig. 58 a–e. Comparison of giant Coster-Kronig (intra-"valence" band) fluctuation and decay processes in atoms **(a)**, molecules **(b, c)** and metals **(d, e)**

In Fig. 58 we compare fluctuation and decay processes, describing dynamic screening and localization, by way of a few schematic examples for atoms, molecules and solids. The atomic (Fig. 58 a) and molecular (Fig. 58 b) cases have been extensively discussed in Sects. 6 and 7. Figure 58 c intended to give a schematic picture of the π-valence band of a polymer-like polyacetylene (π-electrons in a box); intra-band fluctuation processes can in principle shift and broaden the one-electron levels and give rise to shake-up satellites. Figures 58 d, e illustrate the analogous case of dynamic screening and localization of holes in narrow bands in metals. Figures 58 d corresponds to screening of a 3 d-hole in metallic Zn. Hole levels in the ca. 1 eV broad 3 d-band appear to have relaxation shifts as if they were atomic, fully localized holes[149, 174] but there is also clear evidence of band structure[174].

Finally, Fig. 58 e may illustrate intra-3 d-band fluctuation and decay of a 3 $\underline{d}$-valence hole in metallic Ni. This process is probably one of the essential elements in the description of the valence band photoelectron spectrum which seems to show 3 d-band narrowing, decreased splitting of the majority and minority spin bands and a pronounced shake-up structure about 6 eV below the Fermi level (see e.g.[148, 175] and references therein, and also Sect. 8.3.4).

8.2 Model Discussion of Relaxation and Screening of Hole Levels in Extended Systems like Polymers and One-Dimensional Metals

In Sect. 7.1 we discussed dynamic relaxation, localization and screening of core and valence holes in terms of a quasi-hole hopping between the nuclei and followed by an intramolecular screening cloud. If we now create a polymer chain based on a given monomer one might expect to find situations where a hole can start to propagate along the polymer chain; the motion of the quasi-hole must then be associated with a band structure. However, if there are strong intra-monomer relaxation and localization effects, an ordinary one-electron band structure calculation does not work, and the one-electron bands will show relaxation shifts and band narrowing and there could be pronounced molecular-like, shake-up satellite structures. In addition to the intra-monomer dynamic relaxation, since we are dealing with an extended system, there will be charge displacement towards the hole from the entire system, making the quasi-hole appear neutral outside a volume of the order of the unit cell.

Let us look in some detail at the self-energy of a hole in the case of a non-conducting polymer, e.g. polyacetylene, a schematic level structure of which is shown in Fig. 59 a. Following the creation of a core (or inner valence) hole $\underline{c}$, a core electron c′ may fill the primary hole $\underline{c}$ while at the same time exciting a valence electron $v_<$ to empty valence levels $v_>$ above the Fermi level. If conserving energy, this is a Coster-Kronig transition which gives a lifetime to the hole level. However, as discussed in detail in previous sections, one has to consider both virtual and real transitions, i.e. fluctuation and decay, and one is lead to study the fundamental second-order self-energy (Fig. 59 d) as well as different infinite-order renormalizations of it. If we impose the restriction $\underline{c}' = \underline{c}$, not allowing the hole to fluctuate between core level MO's, the core hole will be represented as delocalized over the entire system. For a large system (N atoms) the core hole then becomes associated with vanishing charge (1/N) in each unit cell, and in this approxima-

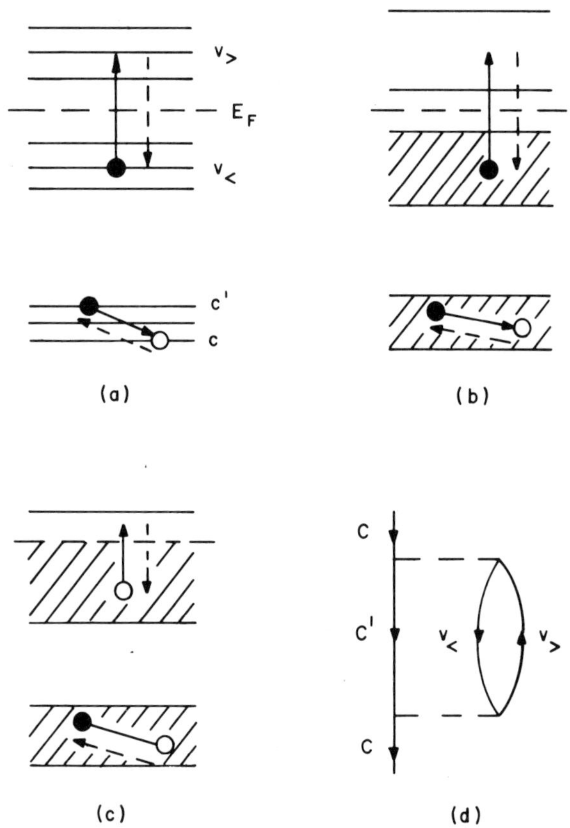

Fig. 59 a–d. Dynamic screening and localization of a core hole in a finite **(a)** and infinite **(b)** polymer and quasi-one-dimensional metal **(c)**; **(d)** shows the lowest order Feynman diagram for the process

tion there is no relaxation. This will happen even for very deep core levels and obviously represents an unphysical restriction when using a delocalized basis.

Fluctuation of the hole between different levels $\underline{c}$, $\underline{c}'$ is equivalent to constructing a wave-packet for the hole based on this band of levels. If we let the hole fluctuate between two consecutive levels we basically mix gerade and ungerade wave functions that differ by one node, forming a wave-packet that is localized to one half of the system. Mixing states from the entire band of levels allows us to localize the hole wave-packet to one particular bond or even one particular atom. In the same way, excitation of valence electrons can mix gerade and ungerade states and localize a valence electron wave-packet to a bond site or an atomic site. The coupled system, as represented by the self-energy, describes the dynamics of screening of the hole by the valence electrons and the propagation of a quasi-hole together with its screening cloud through the polymer. Note, however, that since the screening charge arises from polarization of the valence orbitals there is no net charge transport connected with the screening charge.

In the self-energy (Fig. 59 d), restricting $\underline{c}'$ to the same level (e.g. $2\underline{\sigma}_g$ or $2\underline{\sigma}_u$ in polyacetylene) as the initial hole $\underline{c}$ but allowing fluctuations over all sublevels describes the screening of a hole wave-packet localized to a bond-orbital of a monomer and hopping between monomers. This is probably a reasonable picture for a saturated polymer, having empty σ-orbitals but no π-valence orbitals, and experimental photoelectron spectra for polyethylene and similar systems ([176–178] and references therein) do not show pronounced shake-up structures and level shifts. However, for unsaturated polymers with low-lying empty π-orbitals, the π-electrons are highly polarizable *within* a monomer unit. Temporary localization of the hole wave-packet over an atom within a monomer can then substantially increase the effect of screening and add to the relaxation shift. A quasi-hole wave-packet propagating in the carbon (2s) derived band in e.g. polyacetylene thus should become heavily dressed by the polarizability of the monomer; as a result, the spectral resolution of the quasi-hole will contain excited states in the form of pronounced shake-up satellites in addition to wave-vector dependent shifts of the one-electron bands. In particular, when entering a monomer unit cell a locally $2\underline{\sigma}_g$-like hole may recoil and strongly interact with a locally $2\underline{\sigma}_u 1\underline{\pi}_u 1\pi_g$-like configuration, which might be the origin of the satellite structure observed experimentally in polyacetylene in the 25–30 eV range by Duke et al.[179] using 40.8 eV UPS.

If the band gap could be made to go to zero (Fig. 59c) we would have a quasi one-dimensional metal. In addition to atomic and molecular-like single particle and collective excitations one would then have low-lying excitations characteristic of a metal, namely low-energy electron-hole pairs, giving asymmetric line shapes, and collective plasmon excitations giving rise to satellites. In fact, the inorganic polymer $(SN)_x$ has been found to be a metallic conductor (even a superconductor) with plasmon excitation energies ~ 1.5 eV perpendicular and ~ 2.5 eV (dispersion) parallel to the polymer axis, showing up as pronounced satellite structures on the main $1\underline{s}$ line in the XPS spectrum[180].

8.3 Relaxation, Charge Transfer and Screening of Core Holes in Real Solids

In transition and rare-earth metal compounds the core hole screening process shows up in many spectacular ways due to the presence of low-lying localized and unoccupied levels. Wertheim et al.[158] and Jørgensen and Berthou[159] found very prominent satellites in $3\underline{d}$ XPS spectra of La compounds and Signorelli and Hayes[160], and more recently Suzuki et al.[161] found strong satellites in $4\underline{d}$ spectra. The spectra in Fig. 60 suggest that it may be more relevant to consider the core level as being split into two levels rather than having a satellite. In the presence of a core hole there are two fairly stable and nearly degenerate configurations of the ionic system and which level picks up the most spectral strength in an ionization process depends on the details of the system and may vary for different choices of ligands or metal ions.

The phenomenon can be explained in terms of screening of a core hole on the central metal ion via charge transfer from filled ligand orbitals to empty 4f-levels on the central La ion, as suggested by Wertheim et al.[158] and Jørgensen and Berthou[159]. In the rare-earth compounds the overlap between these orbitals is quite small due to the very strong localization of the 4f-orbitals and it is quite reasonable to argue in terms of atomic orbitals centred on ligands or central metal ions. However, in many other cases there is a considerable overlap and although in the case of the 3d-transition metal compounds one might talk in terms of ligand-to-3d charge-transfer screening, in general it is better to argue in terms of polarization of molecular orbitals. Kim[150], Wallbank et al.[151, 154] and

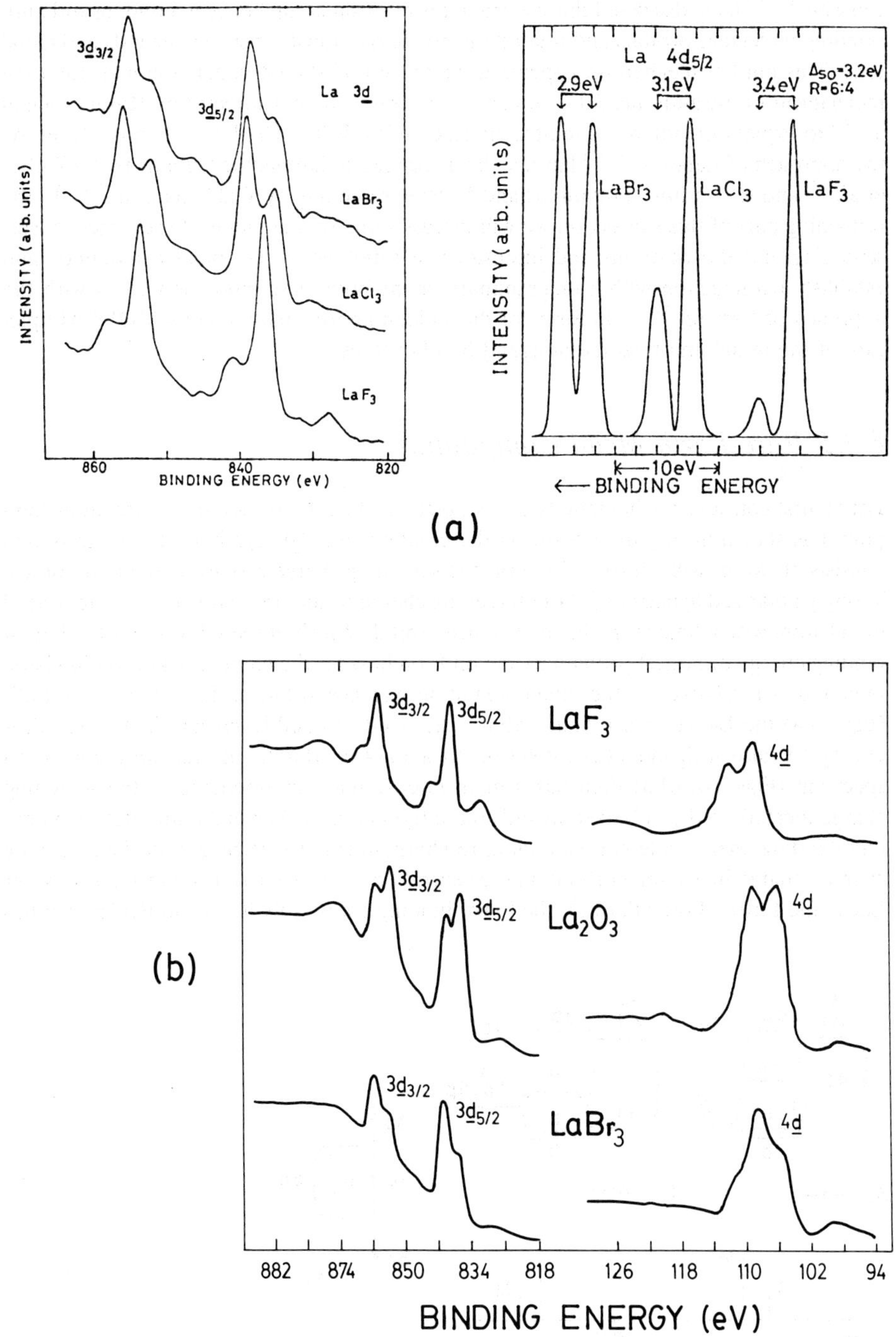

Fig. 60 a, b. X-ray photoelectron spectra of 3 d and 4 d core levels in La compounds; **(a)** from[161]), **(b)** from[160])

Larsson[155–157] have discussed the spectra in terms of shake-up between bonding and antibonding molecular orbitals and related this to charge transfer between ligands and metal ions. This kind of mechanism appears to be more realistic (stronger coupling) than the mechanism of intra-atomic 3 d → 4 s shake-up originally put forward by Rosencwaig et al.[181] to explain the intense satellites on 2 p core levels in 3 d TM compounds. Recently, the approach of Larsson[155–157] has also been applied to lanthanide trifluorides by Weber et al.[163] and to thorium compounds and 5 f-level screening by Wallbank et al.[164]. In the remaining part of this subsection we will discuss some simple models for charge-transfer screening and the positions and intensities of satellites in the metal compounds and establish a connection with a diagrammatic many-body description as well as with the approach of Larsson[155–157] in terms of the sudden approximation and a MOLCAO picture of the initial and final metal-ligand bond orbitals.

8.3.1 Rare-Earth Metal Compounds

Let us first consider a schematic two-level picture of 3 d-ionization of a lanthanum compound as shown in Fig. 61. Following the creation of a 3 d-core hole, the ionic system consists of two possible levels where (a) the screening charge resides on the more or less strongly polarized ligands or (b) the screening charge resides in a 4 f-orbital on the central metal atom with a hole in the ligand valence orbital. Which of these levels has the lowest binding energy or largest probability depends on how far the empty La 4 f-level has been pulled down relative to the filled ligand np-valence orbitals. In the case of LaF_3 (Fig. 61 a) the La 4 f-level remains above the band of filled levels and it then requires energy to raise a ligand F(2 p) electron to a La(4 f) orbital. The resulting core level spectrum (Figs. 60, 61 b) then has a main intense line corresponding to the screening charge located on the polarized ligands (case (i)) and a weaker shake-up satellite on the high-binding energy side corresponding to charge-transfer screening with the screening charge situated in a 4 f-like orbital on the central La ion (case (ii)). If we instead consider $LaCl_3$ and $LaBr_3$ (Figs. 60, 61 c), the lower binding energy of Cl(3 p) and Br(4 p) orbitals

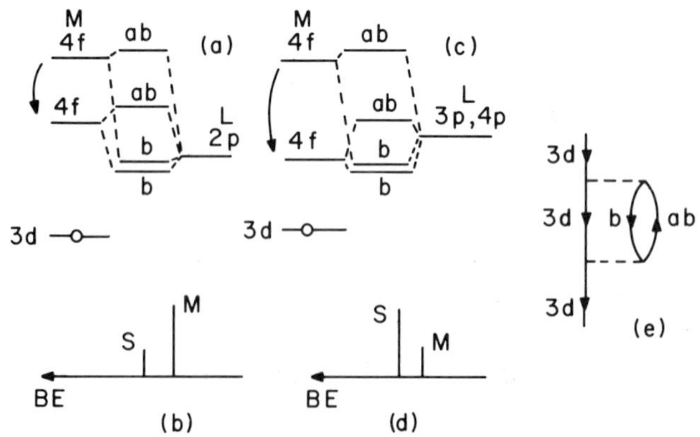

Fig. 61 a–e. Schematic picture of 3 d-ionization in La compounds. LaF_3 (**a, c**), $LaCl_3$ and $LaBr_3$ (**b, d**); (**e**) shows the lowest order Feynman diagram for the relaxation process

relative to F(2 p) (Cl and Br less electronegative than F) seems to cause the La(4 f) level to be pulled down by the 3 d-core hole to below the Br(4 p) valence orbitals (Fig. 61 c). The lowest ionic level (Figs. 60, 61 d) now corresponds to a 3 d-core hole screened by Cl(3 p)/Br(4 p) to La(4 f) charge transfer while the most intense line corresponds to an excited level with the screening charge largely remaining on the ligands. In a two-level situation, the charge-transfer satellite can pick up at most fifty percent of the spectral strength and becomes as intense as the "main" line. This happens when the La(4 f) level is pulled down to become degenerate with the ligand-valence orbital in which case the level splitting is determined by the hopping integral. Such a situation represents covalency: The screening charge is transferred from the ligands to a ligand-metal bond in the final ionic state, and the "shake-up satellite" should correspond to an excitation in the bonding region rather than transfer from ligands to metal or vice versa. E.g. La_2O_3 seems to show such a spectrum. In any other (two-level) case, the charge-transfer line will have less intensity than the non-transfer line, regardless of whether the charge-transfer line appears on the low- or high-binding energy side of the non-transfer line.

Figure 61 e shows the lowest order self-energy diagram, describing relaxation, screening and shake-up. Note that the labels refer to the *initial* state with no 3 d-core hole, in which case b = La(4 f) and ab = F(2 p), Cl(3 p) or Br(4 p). Not only can the core hole cause shake-up but the very nature of the bonding and antibonding orbitals will change under the influence of the core hole (this is of course intimately connected with the shake-up process) (see e.g. Fig. 61 a). We shall discuss this in somewhat greater detail in Sect. 8.3.3.

A semi-quantitative description of the core level spectrum and the charge-transfer process can be obtained from a simple two-level MOLCAO-model based on the sudden approximation[155–157, 160]. Here, we follow the formulation of Larsson[157] and consider the influence of a core hole on a single electron in an MO formed by linear combination of AO's u_L and u_M centred on the ligands (L) and the central metal ion (M). In the ground state, before ionization, the electron is in a bonding orbital

$$\phi_i^b(N) = u_L \cos\eta + u_M \sin\eta \tag{86}$$

where the antibonding counterpart is empty. Neglecting relaxation effects, the N-1 electron ionic ground state will be described by the same MO as in Eq. (86) (frozen orbital approximation)

$$\phi_i(N-1) = u_L \cos\eta + u_M \sin\eta \tag{87}$$

(although we keep the index i, this orbital refers to the *final* state). The true bonding (b) and antibonding (ab) ionic levels can be written as

$$\phi_f^b(N-1) = u_L \cos\xi + u_M \sin\xi \tag{88}$$

$$\phi_f^{ab}(N-1) = u_L \sin\xi - u_M \cos\xi \tag{89}$$

and combining Eqs. (87)–(89) one obtains

$$\phi_i(N-1) = \phi_f^b(N-1)\cos(\xi-\eta) + \phi_f^{ab}(N-1)\sin(\xi-\eta) \tag{90}$$

which gives the components of the frozen MO projected on the eigenstates of the system after ionization. In the sudden approximation the probability of reaching the bonding and antibonding final state levels is given by the square of the modulus of the expansion coefficients, and the core level intensity ratio becomes

$$\frac{I^{ab}}{I^{b}} \equiv \frac{I_S}{I_M} = \frac{\sin^2(\xi - \eta)}{\cos^2(\xi - \eta)} = \tan^2(\xi - \eta) \tag{91}$$

In this model the intensity ratio depends on the angle of rotation $\xi - \eta$ due to the perturbing core hole and not on the absolute angles (mixing coefficients) ξ and η.

As an example, let us apply this model to 3d-ionization of La-halides (M = 4f, L = 2p(F), 3p(Cl), 4p(Br)). A reasonable approximation consists in taking the metal and ligand atomic orbitals to be localized and the 4f-metal orbital to be empty in the ground state. This is achieved by choosing $\eta = 0$, and means that instantly after ionization the screening charge is localized on the ligands,

$$\phi_i(N - 1) \simeq u_L \tag{92}$$

In terms of the true core level spectrum, the intensity ratio now becomes

$$\frac{I_S}{I_M} = \frac{\sin^2\xi}{\cos^2\xi} \tag{93}$$

The values of the mixing coefficients $\cos\xi$ and $\sin\xi$ depend on the coupling strength, which we can define in terms of the resonance (hopping) integral H_{12} measured relative to the distance H_{11}–H_{22} between AO levels, giving[157)]

$$\tan 2\xi = \frac{2H_{12}}{H_{11}-H_{22}} \tag{94}$$

If we assume for simplicity that H_{12} is the same for all of the halogen ligands, the core level spectra can be characterized in terms of the relative positions of the metal 4f-level and the ligand valence levels in the different compounds. In LaF_3, after ionization, the empty metal 4f-level appears to remain well above the filled ligand 2p-valence levels so that the distance between the bonding and antibonding levels is largely determined by H_{11}–H_{22} and not by $2H_{12}$. As a result, the angle ξ is relatively small and the satellite-to-main line intensity ratio becomes much smaller than unity. In $LaCl_3$ and $LaBr_3$ it seems that the metal 4f-level is pulled below the top of the ligand valence band. The lesser electronegativity of Cl and Br thus leads to a degeneracy and strong coupling in the final state and to a satellite-to-main line intensity ratio of the order of one. Comparison with experiment actually suggests that the intensity ratio is considerably larger than unity[160, 161)] (Fig. 64) so that the satellite is more intense than the main line. In principle, one can imagine the extreme case that the main line becomes much weaker than the satellite line but this does not seem to have been observed experimentally. The use of LaI_3 might represent a step towards such a situation. Mathematically, this case is characterized by $\eta = 0$ in the initial state and $\xi = \pi/2$ in the final state, resulting in a vanishing overlap between the frozen initial state and the true ionic ground state

$$\phi_i(N-1) \simeq u_L \tag{95 a}$$

$$\phi_f^b(N-1) \simeq u_M \tag{95 b}$$

$$\phi_f^{ab}(N-1) \simeq u_L \tag{95 c}$$

Formulated in another way, the metal (M)-level is suddenly pulled down from well above to well below the ligand (L) valence levels a distance larger than the exchange splitting $2H_{12}$; as a consequence, the system cannot follow adiabatically but undergoes a diabatic (curve crossing) transition to an excited state of the ionic system.

In the above examples the ground-state 4f-levels were initially empty. However, according to Weber et al.[163], YbF_3 respresents the opposite case in that the 4f-levels are slightly below the ligand 2p valence levels already in the ground state so that $\eta > \pi/4$ in Eq. (87). Switching on the core hole now pulls the 4f-levels below the valence band, making the bonding orbital almost pure M(4f) and the "anti-bonding" orbital almost pure L(2p). This corresponds to $\xi \simeq \pi/2$ in Eqs. (88) and (89) and results in

$$\frac{I_S}{I_M} \leqslant \frac{\sin^2(\pi/2 - \pi/4)}{\cos^2(\pi/2 - \pi/4)} = 1 \tag{96}$$

The intensity distribution between the main line and the satellite line resembles the case of LaF_3 but there are however some very important differences regarding characterization of the peaks. In LaF_3 (Fig. 61) the main line corresponds to a strongly polarized ligand valence charge and the shake-up satellite to ligand-to-metal transfer of the screening charge to a 4f-orbital localized on the central La ion. In YbF_3, on the other hand, the main (and also most intense) line corresponds to all of the metal 4f-levels being filled and the satellite then (at least in this model) corresponds to metal-*to-ligand* charge transfer[163].

8.3.2 Rare-Earth Metals

The extreme case of a very weak lowest main line and a strong satellite does not seem to occur (or at least has not been observed) in any REM complexes but it does seem to exist in the 3d-core level spectrum of rare-earth metals and alloys[10, 172, 173, 182]. In La to Nd metal XPS spectra in Fig. 62, the lowest ionic level occurs as a bump on the low-binding energy side of the dominant 3d core line. Just as in the case of La compounds one may characterize the situation in terms of an atomic-like localized 4f-level being pulled down into the metal 6s/5d valence band[14, 24, 145, 173, 183–184]. However, the coupling between the 4f and valence levels appears to be a good deal weaker in the metal than in the compounds (there are probably a number of reasons such as symmetry and selection rules, the 6s/5d band being fairly broad, and screening of the interaction). The final ionic *ground* state therefore must be nearly orthogonal to the frozen ionic state and we then have the situation in Eq. (95). In other words, the probability of filling the 4f level is very low and the main intensity will go into a core line corresponding to an excited state where the 4f-screening orbital is empty. Nevertheless, this core hole has to be screened out within a distance of the order of the lattice spacing and here the 5d-electrons may come

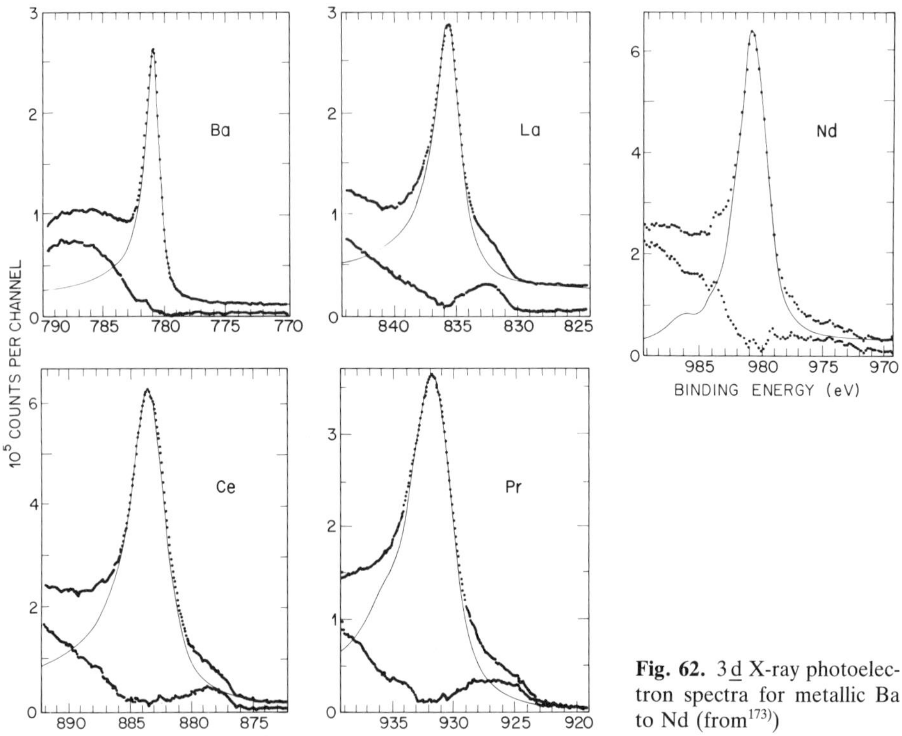

Fig. 62. 3 d X-ray photoelectron spectra for metallic Ba to Nd (from[173])

into play. The screening cloud should resemble an atomic, localized 5 d orbital in the sense of increasing the 5 d character inside the unit cell corresponding to one unit of electronic charge. One may then interpret the weak, lowest (main) line in terms of a 3 d core hole with screening charge occupying 4 f-orbitals, and the intense higher binding energy line in terms of a screening cloud with a pronounced localized 5 d character and the 4 f-orbitals being empty.

One may now ask what kind of spectral features can be associated with the case that all of the more or less localized screening orbitals (or scattering resonances) are empty. A distinct possibility would be shake-up from, loosely speaking, the 5 d-like screening cloud to empty 5 d and other levels above the Fermi level, forming a two-hole-one-electron excited configuration. The Coulomb interaction between the holes may lead to substantial localization of the 5 d-hole and to a lowering of the energy of the double vacancy by perhaps 10–15 eV (the 3 d–5 d interaction should be much stronger than the 5 d-band width so that discrete 3 d5 d levels may be split off from the band. Since the shake-up in this model would have essentially intra-5 d-band character, matrix elements would be large and it is tempting to interpret the prominent, broad satellite structure 10–15 eV above the major 3 d-core lines in the La and Ce metal XPS spectra[10] in terms of 5 d-intra-band shake-up, strongly modified and localized by the presence of the 3 d core hole. This is in line with current ideas about the 3 d-transition metals, as will be further discussed in Sect. 8.3.4 (see also e.g.[175]).

8.3.3 3 d-Transition Metal Compounds

In the 3 d-transition metal compounds there is an appreciable overlap between the metal-3 d and the ligand valence orbitals, resulting in fairly large bonding-antibonding splitting, of the order of 5–6 eV. Although this is still a rather narrow band, it is considerably wider than in the 4 f-REM case, and one can probably no longer even approximately argue in terms of charge transfer between localized atomic-like orbitals[156)]. There is now a strong covalency and the charge-transfer process should then more have the character of displacement of bonding charge. Also, the distinction between inter-atomic charge transfer and intra-atomic shake-up should become less clear. The problem then becomes related to that of core level structure in molecules like N_2 and CO (Sect. 7): In the same way as part of the relaxation and shake-up processes could be discussed in terms of $\pi \rightarrow \pi^*$ and $\sigma \rightarrow \sigma^*$ transitions, in the case of 3 d-metal compounds we shall consider $e \rightarrow e^*$ and $t_2 \rightarrow t_2^*$ transitions (in tetrahedral symmetry; in octahedral symmetry this become $e_g \rightarrow e_g^*$ and $t_{2g} \rightarrow t_{2g}^*$)[150, 154–157, 185)]. This immediately suggests that many-body effects might be important, and a number of the relevant self-energy diagrams for a 2 p core hole are shown in Fig. 63. The fundamental second-order diagrams in Figs. 63 a, b describe relaxation and shake-up in a one-electron sense: Although displacement of a particular electron from the ligands towards the metal ion does set up an electric field, this is not felt by the other electrons in the e and t_2 bonds. In order to include such effects in a many-body framework one has to include the infinite sequence of diagrams in Fig. 63 c, describing the random-phase approximation with exchange (RPAE) and the dielectric response of the bonding charge. Other types of probably important many-electron effects are associated with interactions between core and valence holes and localization of valence holes. Figs. 63 d, e describe to lowest order the energy shift of the valence orbitals, while Figs. 63 f, g illustrate polarization of the valence orbitals due to the core hole. It should be noted that higher order interference between Figs. 63 d, e and Figs. 63 f, g might be important: Strong relaxation of the valence orbitals might considerably influence the core-valence hole-hole and hole-particle interactions. Finally, Fig. 63 h is absolutely essential for obtaining correct satellite-to-main-line distances and Fig. 63 i is probably very important for a correct description of satellite intensities (cf. Sect. 7.2).

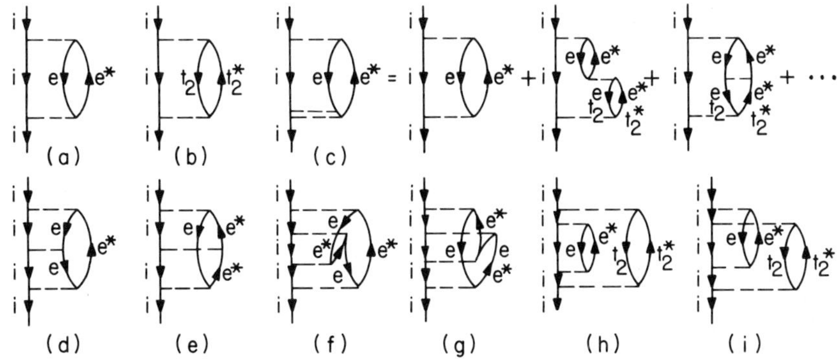

Fig. 63 a–i. Core hole self-energy diagrams describing relaxations in a transition-metal compound (see text)

Since we are discussing cases where the core hole is localized on the central ion and does not break the molecular symmetry, it would seem that a ΔSCF calculation should be able to account for the energy separation between the main line and the shake-up satellites. In particular, such a calculation should effectively incorporate the diagrams in Fig. 63 to infinite order. However, so far, calculations have only been performed within an SCF-MSXα framework ([155–157, 185, 186], and references therein), and although satellite-to-main-line intensity ratios are often reasonably well described, satellite-to-main-line energy separations systematically come out too small, sometimes much too small[156]. This has led Tossel[185] to question the relevance of the charge-transfer model for describing the satellites, and to propose a mechanism involving metal (3d) to conduction band transitions. However, most likely, the corresponding transition matrix elements are too small to account for the satellite intensity[156, 186] and the problem might rather lie with the SCF-MSXα calculation itself. It seems that the satellite excitation energy is taken to be the energy splitting between the bonding-antibonding orbitals in the final ionic *ground* state rather than between the $2\underline{p}$ and $2\underline{p}\,\underline{e}\,e^*$, $2\underline{p}\,\underline{t_2}\,t_2^*$ SCF configurations, and it cannot be excluded that Coulomb interactions within the satellite configuration (Figs. 63 d–g) can lead to a significant increase of the excitation energy.

It seems to be far too early to give a systematic account of the development of satellite-to-main line intensities and energy separations along the sequence of transition metals as well as for different compounds of a given 3d metal. For one thing systematic experimental information is somewhat lacking and, when existing, sometimes seems confusing. For example, the collection of experimental $2\underline{p}_{3/2}$ XPS spectra of Vernon et al.[152] suggests that for compounds of Sc to Cr the energy separation ΔE and intensity ratio I_S/I_M is $\sim$11–14 eV and $\sim$0.1–0.25 resp., and there does not seem to be any systematic variation with ligand character. In particular, ScF_3, $ScCl_3$ and $ScBr_3$ have $\Delta E \simeq 12$ eV with $I_S/I_M = \sim 0.25$, ~ 0 and ~ 0.1 resp. On the other hand, the sequence TiF_4, $TiCl_4$, $TiBr_4$ and TiI_4[154] shows ΔE decreasing smoothly from 13 to 7 eV and I_S/I_M slowly increasing. A very interesting jump in ΔE and I_S/I_M occurs between Cr and Mn compounds, ΔE decreasing and I_S/I_M increasing roughly by a factor of two, and proceeding towards Co and Ni this trend is further amplified. Finally, in Cu(II) compounds ($3d^9$ configuration), there is a slight increase in ΔE and decrease in I_S/I_M and the dependence on the ligand electronegativity has become reversed.

In spite of possible exceptions and deviations for particular systems, the general conclusions of Larsson[155–157] (see also Kim[150]) are probably valid which would lead to the following approximate characterization of 3d-transition metal-compound core level spectra:

a) Sc, Ti compounds: In the final ionic state the M(3d) levels are still well above the L(np) valence levels which directly results in fairly large ΔE and small to moderate I_S/I_M values. If there are no further complications ΔE should decrease and I_S/I_M increase with declining electronegativity of the ligands. TiF_4 to TiI_4 nicely follow such a trend[154] but otherwise there is no clear picture available[152]. Since the final ionic state is rather covalent, the main $2\underline{p}$ XPS peaks (Fig. 64) should be associated with substantial charge transfer from ligand to bonding region, and the shake-up satellite should correspond to further charge transfer towards the metal ion, or perhaps rather away from the bonding region. This characterization should have some applicability also to V and Cr compounds, although the open 3d-shell may cause a number of complications. Note that intensity ratios $I_S/I_M \simeq 0.25$ at energy separations $\Delta E \simeq 12$ eV are signs of very strong

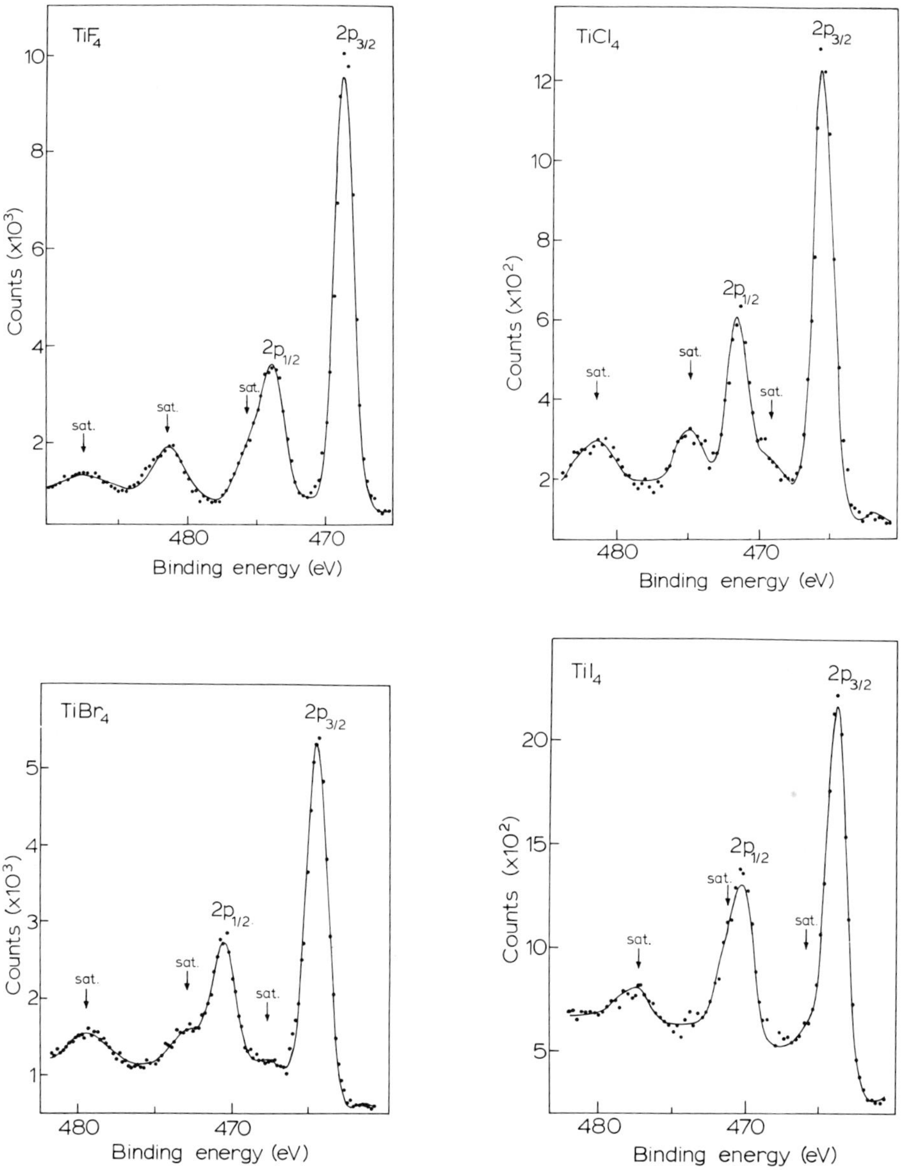

Fig. 64. 2p X-ray photoelectron spectra of TiX_4 compounds (from[154])

coupling, as will be evident below when the satellite separation ΔE becomes smaller. b) Mn to Ni compounds: Experiments[152] (Fig. 65) show that the satellite-to-main line separation ΔE has dropped to the 4–7 eV range and the intensity ratio I_S/I_M is 2–4 times as large in comparison with (a). Roughly speaking, it seems that the empty 3d-levels have been pulled down to about the same energy as the ligand p-levels so that the excitation energy is of the order of the bonding-antibonding splitting. Since the antibonding levels now are becoming filled, the collective response of the bond will decrease and

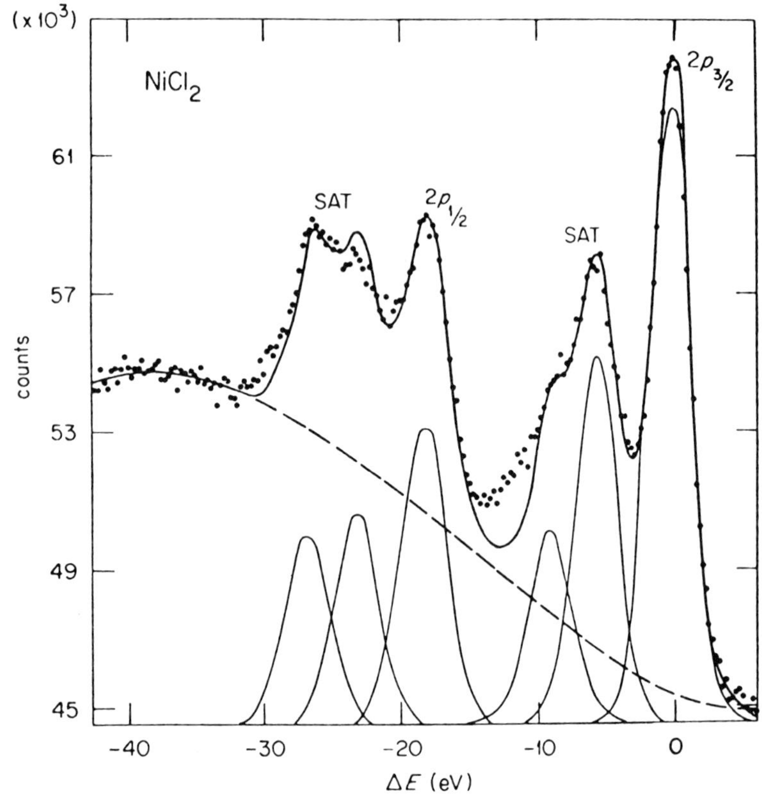

Fig. 65. 2 p X-ray photoelectron spectrum of $NiCl_2$[152)]

this could possibly be one reason for the slow decrease of ΔE along the series of metal compounds. The jump between Cr and Mn might have to do with contraction of 3 d-orbitals and decrease of overlap and bonding-antibonding splitting. Since in Mn to Ni compounds we are going from a neutral ground state with large ionicity to a final ionic ground state with strong covalency, the intensity of the satellite can become comparable to that of the main line. Again it seems reasonable to associate the main line with charge transfer from the ligands to the bonding region and the shake-up satellite with charge transfer away from the bonding region.

c) Cu(II) compounds: Here one e*-orbital remains empty and there is one possible e → e* transition. Calculations by Larsson[155)] suggest that the neutral ground state is strongly covalent while the final ionic ground state is strongly ionic with the Cu(3 d) levels pulled way below the valence levels. The main line then corresponds to a $3\,d^{10}$ configuration where charge transfer from the bonding region to an empty 3 d-orbital on the Cu ion has occurred, and the shake-up satellite then must correspond to Cu-to-ligand charge transfer[155)]. This is a situation which leads to strong satellites (cf. Eq. (96)), in agreement with experiment[10, 152)]. In this case, declining electronegativity will lead to *increasing* ΔE and decreasing intensity ratio I_S/I_M, again in agreement with experiment[152)]. Finally, in view of the above discussion, we may characterize the Cu(II) main 2 p line in terms of a valence hole localized on the ligand and the shake-up satellite in terms of a core hole and a 3 d-hole localized on the metal ion.

8.3.4 3d-Transition Metals

In analogy with the 3d-TM compounds, core level XPS spectra of the metals show satellite structures in the 5–15 eV range above the main lines. However, in contrast to the compounds, in the metals the satellite intensities are always much weaker than the corresponding main line intensities[10, 175, 187–191] and references therein. The problem of localization of screening charges and shake-up satellites in core hole spectra in narrow band metals has attracted much theoretical attention[47, 132, 145, 175, 192, 193] and references therein, but although model calculations have given much insight into the dynamics of the screening process, a quantitative description of real spectra from first principles is not yet feasible in general.

Ni metal is a particularly interesting system, showing what seems to be pronounced effects of localization. Hüfner and Wertheim[189] have noted that the lack of multiplet splitting of the main $3\underline{s}$ XPS line in Ni implies that the local, screened, ionic configuration must be $3\underline{s}^1 3d^{10} 4s^1$ instead of $3s^1 3d^9 4s^1$ plus screening charge. The perturbation thus leads to a *locally filled 3d-band* and the ionic ground state can be viewed as a core hole surrounded by a localized screening cloud with atomic 3d character. If the 3d-band becomes filled locally, it needs no longer be locally pinned to the Fermi level and a localized level could actually be split off from the bottom of the 3d-band, as found by Davis and Feldkamp within a Hubbard model[193]. However, this is precisely a situation where there could be appreciable probability of finding the split-off level and the localized 3d-screening cloud unoccupied (cf. Cu(II) compounds, Sect. 8.3.3, and Eq. (96)). In this shake-up state there is a localized aggregate of a core hole and a $3\underline{d}$-hole, with the shake-up electron in extended states above the Fermi level or perhaps in a semi-localized state[175, 193] and references therein.

The localized atomic-like character of a $3\underline{p}\,3\underline{d}$ double vacancy in Ni metal shows up very clearly in the form of atomic multiplet structures in Auger spectra[10, 194]. The really interesting thing, however, is that also Auger spectra with a $3\underline{d}^2$ double vacancy in the final state show clear atomic-like effects[10, 194] and cannot be described within a one-electron band picture for the $3\underline{d}$-holes. Including Coulomb repulsion between the holes within a narrow-band Hubbard model, one finds[194–198] that bound localized two-hole levels can become split off from the valence band. This immediately forms a basis for shake-up satellites analogous to those found on core levels[187] and also suggests that the dynamics of a hole in the 3d-valence band might be strongly modified by coupling to localized two-hole-one-electron excitations. The corresponding lowest order self-energy diagram for a $3\underline{d}$-hole is shown in Fig. 66a. It represents an intra-3d-band (giant Coster-

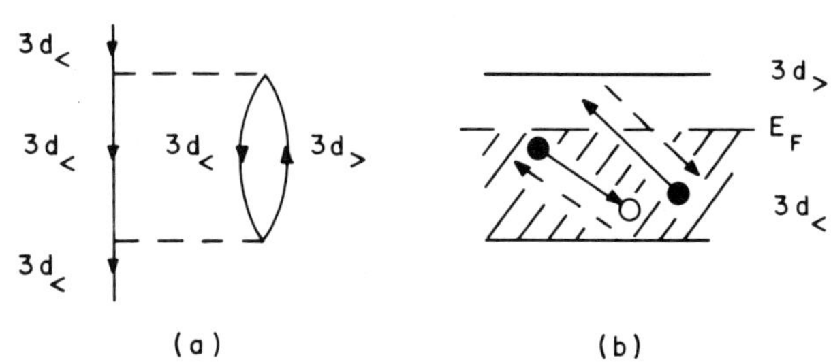

Fig. 66a, b. Intra-3d-valence band (giant Coster-Kronig) fluctuation and decay of a valence hole

Kronig) fluctuation and decay process (Fig. 66 b) and describes the propagation of a 3 d-hole wave packet together with its 3 d-like screening cloud. In the case of Ni metal, the intermediate two-hole-one-electron excitations contain strongly localized states which lead to strong energy variations of the self-energy and to energy shifts of the one-electron bands and to shake-up levels[198, 199]. The same kind of results have been obtained by Davis and Feldkamp[193, 200] using a narrow-band Hubbard model (for more details regarding the Ni 3 d-problem and for further references see[175] and references therein).

The 3 d-transition metals plus a number of the following elements from Cu to, say, Rb metal offer a perhaps unique opportunity to study the problem of how a localized hole in a deep core level develops into a propagating hole in the valence band of a metal. In Rb metal (Z=37) the 3 d-level is deep and the conduction electrons are free-electron-like. In Kr (Z=36) the 3 d-level is still deep but the system is a rare gas atom and provides the free atom aspects of relaxation. Going to lower atomic number Z, the 3 d-binding energy decreases and the 3 d-radius increases, and in metallic Zn a narrow band has developed[149, 174], with ~1 eV band width and with a clearly observable band structure and band dispersion[174]. However, with this small band width, a 3 d-hole wave packet will move very slowly and will be screened nearly as well as a localized hole. The associated relaxation shift is of the order of 5 eV and it follows that a band structure calculation cannot place the 3 d-bands at the right energy[149]. In Cu, the 3 d-bands are ~3 eV wide and the top of the 3 d-band is ~2 eV below the Fermi level. Here, it seems that a one-electron band picture applies very well; the 3 d-band is completely filled and intra-3 d-band fluctuation and localization cannot take place. On the other hand, this seems to be precisely what happens in metallic Ni, which very nicely illustrates that it is not the band width per se which is important but the band width relative to the interaction strength. Proceeding towards Co and Fe, finally, the results of Eastman et al.[175] suggest that a one-electron picture becomes increasingly good again, although there seem to be significant effects of localization and level shifts also in Co. This seems reasonable since Co Auger spectra with $3\underline{d}^2$ final states do not show atomic-like behaviour; a double vacancy in the 3 d-band is thus much less localized than in Ni and consequently there will be less localization of single 3 d-holes. Generally speaking, one can probably argue that the higher number of vacancies in the 3 d-bands in the ground state makes the screening of the hole-hole interaction much more effective in Fe and Co than in Ni, leading to small or no band narrowing and no shake-up satellites (at least, so far not observed). However, to date no quantitative investigations of these types of problems seem to have been performed.

8.4 Relaxation and Shake-Up in Adsorbate Systems

As a final example of systems showing charge-transfer relaxation following photoionization, we shall discuss screening and shake-up associated with the creation of a core hole in an atom or molecule adsorbed on a metal surface or a metal cluster, or in a ligand molecule in a transition-metal complex. For a discussion of the experimental situation and for further references, we refer to some recent papers by Fuggle et al.[201] (CO and N_2 adsorbed on TM-surfaces), Plummer et al.[171] (solid and gas-phase carbonyls) and Bancroft et al.[202] (gas-phase TM hexacarbonyls). Experiment[171] shows that photoelectron spectra of multimetal carbonyls and CO adsorbed on a corresponding TM-surface are quantitatively very similar, and that there is qualitative agreement with spectra for single-metal carbonyls. This suggests that the chemisorption problem can be studied from a number of different angles, and that purely molecular calculations may be of great importance for understanding the nature of substrate-adsorbate charge-transfer processes and the chemisorption bond.

On the theoretical side, a number of large-scale, ab initio SCF calculations have been performed in order to study the screening of a core hole on an adsorbed atom or molecule[165, 166, 203, 204]. In the case of Na, Si and Cl adsorbed on jellium (semi-infinite electron gas)[165] or on an Al_5 cluster[166], Lang and Williams[165] and Hoogewijs and Vennik[166] conclude that in the ionic ground state, screening occurs through charge transfer from the substrate to localized, atomic-like orbitals on the adatom. This means that the screening charge mainly resides on the adatom and cannot be described in terms of surface plasmons and screening charge localized on the metal surface. Ellis et al.[203] and Baerends and Ros[204] have studied the case of CO adsorbed on a Ni_5 cluster and found that C(1s) and O(1s) holes become screened through transfer of one unit of charge to the low-lying, almost empty 2π orbitals on CO, i.e. through the back-bonding channel.

The above-mentioned ΔSCF calculations are concerned with charge-transfer processes leading to the ionic ground state and give information on the binding energy shift due to both intra- and extra-molecular relaxation. Also, with knowledge of wave functions and overlaps, one can get information about the probability of reaching the ionic ground

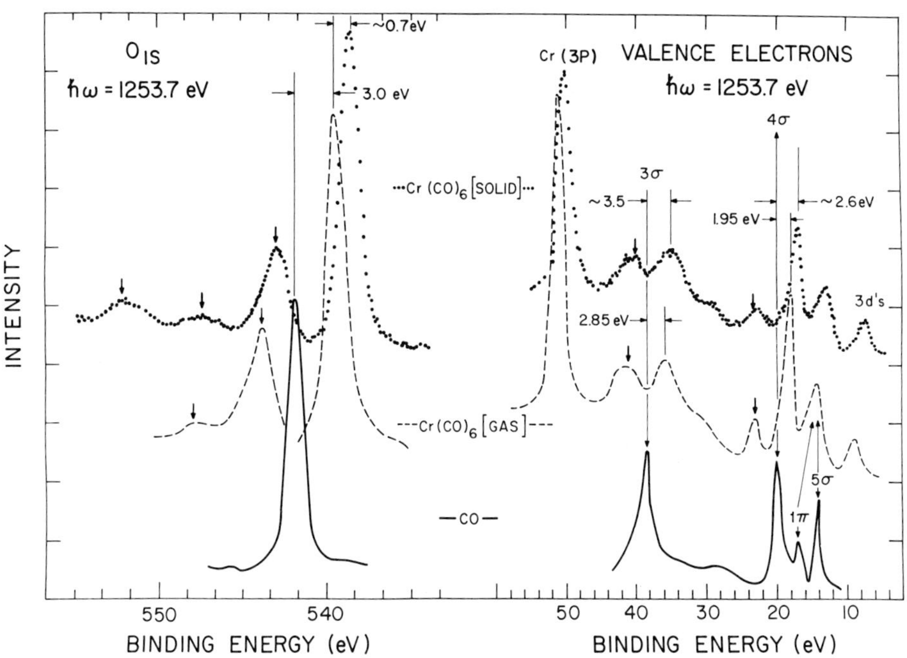

Fig. 67. X-ray photoelectron spectra of metal-carbonyls (from[171])

state, giving the intensity of the main photoelectron line relative to the total intensity of shake-up and shake-off processes (cf. Sect. 8.3.1). However, the presence of very strong satellite lines in C(1 s) and O(1 s) core level spectra for CO adsorbed on TM-surfaces or in TM-carbonyl complexes and solids (Fig. 67) is a clear indication that the system has high probability of being left in an excited ionic state. At present, large interest is focussed on the nature of these excited states and what they may tell about charge-transfer processes and the bonding of chemisorbed atoms and molecules ([147, 167–171, 201, 202, 205, 206] and references cited therein).

In order to interpret X-ray photoelectron spectra with main lines and shake-up satellites, one has to study the dynamics of the screening process following the sudden creation of a core hole. This can be done in at least three different ways with many variations:

(a) In a time-dependent picture, solving the equation of motion for a core hole coupled to the adsorbate and substrate, and then making a Fourier transformation to energy space to obtain the intensity distribution over energy eigenstates;

(b) using e.g. diagrammatic many-body theory in energy and configuration space, calculating the core hole self-energy and spectral function on a basis suitable for the system to be studied (cf. Sects. 5–7);

(c) in a time-independent picture, calculating ground and excited states of the total ionic system and projecting onto the frozen ionic system (electronic and nuclear configuration of the initial state) to obtain the final state probability distribution (cf. Sect. 8.3.1).

In any of these methods, the system itself can be described with varying degrees of sophistication, from simple models to extensive ab initio multi-configuration SCF descriptions.

So far, the chemisorption problem has mainly been discussed in terms of model calculations[147, 167–170] using methods (a) and (b) above. However, Gunnarsson and Schönhammer[169] have also demonstrated that realistic model calculations, taking into account the structure of the substrate and adsorbate by a comparison with band structures and molecular calculations, can give very good descriptions of core level spectra of CO and N_2 adsorbed on Ni and Cu metal surfaces. Accurate basis sets have not yet been employed in conjuntion with the many-body theory to describe charge transfer and shake-up in adsorbate and metal-carbonyl or nitrosyl systems. However, Domcke et al.[118, 206] have studied the similar problem of photoionization of the NO_2 and NH_2 groups in para-nitroaniline, p-$NO_2C_6H_4NH_2$, and have indeed found examples of strong charge-transfer processes where a core-level (NO_2 1 s) satellite line becomes as intense as the main line (cf. Sect. 8.3.1).

Method (c), finally, has recently been used by Loubriel[205] to describe relaxation and shake-up in single $Ni(CO)_4$ and $Cr(NO)_4$ molecules within an MSXα framework. According to Loubriel, the most prominent shake-up satellite at around 6 eV above the main line (Fig. 67)[207] in $Ni(CO)_4$, as well as in e.g. $M(CO)_6$ with M=Cr, Mo or W, is basically due to intramolecular $1\pi \rightarrow 2\pi$ shake-up on CO, instead of metal $\rightarrow$ CO(2π) charge transfer, which has been the usual interpretation ([169, 171, 202, 204], and references therein). Evidently, there are two conflicting views, and I shall conclude this discussion by producing a list of facts and conclusions which might give a guideline to possible solutions to the problem.

1) Accurate LCAO-SCF-Xα calculations (no muffin-tin approximation) for a Ni_5:CO cluster[203, 204] give the following picture:

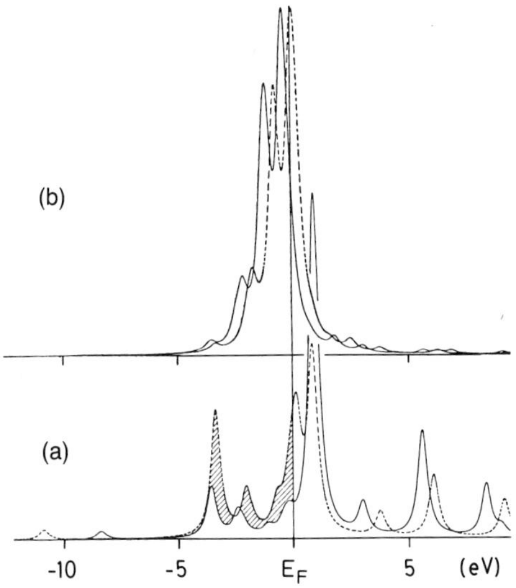

Fig. 68 a, b. Local density-of-states plot in Ni_5: CO. **(a)** CO-2π orbital, **(b)** Ni-3 d orbital; in the ground state of Ni_5: CO, *solid line*, in the core hole state Ni_5: CO^+, *dashed line* (from[204])

(a) Neutral ground state: CO(2π) hybridized with the Ni substrate, enough 2π character below the Fermi level to allow charge transfer of ~ 0.5e from the substrate to 2π levels localized on the CO molecule (back donation) (Figs. 68 a, b).

(b) With a core hole localized on CO, the 2π levels become pulled down enough to allow additional charge transfer of ~ 0.9e to the 2π levels (Figs. 68 a, b). Therefore, in the ionic ground state the screening charge is localized on the CO molecule with the core hole, and the screening charge goes into 2π-like orbitals.

(c) The muffin-tin approximation leads to a poor description of the energy level structure in the valence region (see also[208]).

2) Loubriel[205] uses a SCF–MSXα approach with muffin-tin approximation to study a $Ni(CO)_4$ molecule, reaching the following conclusions:

(a) Neutral ground state: The amount of 2π-mixing is found to be about the same as in the presumably more accurate non-muffin-tin calculations mentioned above. The picture of CO–Ni bonding therefore appears to be equivalent to that in 1(a).

(b) Ionic ground state, $1\underline{s}$ core hole localized on a carbon atom in a particular CO molecule: There is an inflow of 1e of screening charge into the carbon MT (muffin-tin) sphere having the $1\underline{s}$ hole, with about 0.6e due to the CO(1π) orbitals and most of the remaining part coming from 4σ and 3σ, also localized on CO.

3) Comment on points 1) and 2): It seems reasonable, in my opinion, to consider the non-MT results[203, 204] as a reference, representing the state of the art. Also, assuming that the cases of Ni_5 : CO and $Ni(CO)_4$ are not fundamentally different, the following comments can be made:

The physical picture of screening of a C($1\underline{s}$) hole should be a considerable redistribution of the localized CO(1π) orbitals (just as for free CO) *together with* transfer of 1e of screening charge into CO(2π) orbitals from other parts of the molecule or cluster. Evidently, the redistribution of the CO orbitals does not lead to long-range screening: This can only be done via charge transfer from the substrate to the CO(2π) orbitals. Strictly speaking, this also represents a redistribution of localized orbitals: True long-range order can only exist in an infinite system. Regarding the calculation of Loubriel[205], it could be that the MSXα scheme provides a basically correct picture of the screening process but that the analysis of the charge flow into the carbon MT sphere only gives part of the picture, not revealing the proper nature of the 2π levels.

We have now come to the very difficult problem of *ionic excited states and shake-up satellites,* and we can continue the list of facts and fiction in the following way:

4) A characterization of the shake-up excitations has to be based on the charge distribution of the final *ionic* excited states.

5) Assuming that the metal $\rightarrow$ CO(2π) charge transfer dominates the screening of the hole, making the CO molecular ion appear neutral relative to the substrate, does the prominent 6 eV shake-up satellite involve (a) metal $\rightarrow$ CO(2π) charge transfer, (b) charge transfer back to the substrate or (c) excitation localized on CO or in the bonding region?

Process (a) has been suggested by Bancroft et al.[202] and many other workers before them (see[202] and references cited therein). This interpretation[202] is based on MSXα calculations of the level structure of metal hexacarbonyls using the *neutral* ground-state potential.

Process (b) describes the consequences of models where the adsorbate 2π levels are considered to be either filled or empty, as discussed e.g. by Gunnarsson and Schönhammer[147, 169], Gumhalter[170], Plummer et al.[171] and Baerends and Ros[203]. Relative to the ionic ground state, an empty 2π screening orbital represents shake-up and charge transfer back to the substrate, while relative to the initial ground state it represents no charge transfer to the adsorbate molecular ion.

Process (c) could be thought of as various excited states of the screening charge localized in the region of the adsorbate, the "surface molecule", and the problem becomes more that of a molecule embedded in an electron gas[167, 168]. The satellite could e.g. be due to shake-up from the metal-CO(2π) bonding levels to CO(2π) levels still more localized on CO. Another interesting possibility would be internal shake-up of CO, e.g. $1\pi \rightarrow 2\pi$, as advocated by Loubriel[205]. Since this leads to an ionic CO molecule, screening and perhaps charge transfer to CO will have to take place simultaneously. In Loubriel's calculation on $Ni(CO)_4$ this is automatically taken care of, within the capacity of the MSXα scheme, since all orbitals are calculated in the presence of the C($1\underline{s}$) hole.

6) Comments on point 5 above: In view of the fact that the 6 eV satellite is particularly prominent and well-defined in molecular and solid carbonyls, it seems reasonable to discuss the origin of the satellites within a molecular framework. The following comments and speculations might then be of some relevance:

a) In Loubriel's[205] calculation, the lowest empty level in the presence of a C($1\underline{s}$) hole is a 2π level localized on CO (Fig. 68). Likewise, the cluster calculations suggest a high density of empty 2π-like orbitals just above the Fermi level in the presence of the hole[204] (Fig. 68 a). It seems difficult to avoid the conclusion that these empty 2π orbitals will be final states for much of the prominent shake-structure.

b) In terms of the level structure of the *final ionic state,* the following configurations appear to be obvious candidates (mixed notation: C($1\underline{s}$) = CO($2\underline{\sigma}$), b = metal(3d)–CO(2π) bonding orbital, $2\pi^*$= empty 2π orbital localized on CO)

$$2\underline{\sigma}\,\underline{b}\,2\pi^* \tag{97 a}$$

$$2\underline{\sigma}\,1\underline{\pi}\,2\pi^* \tag{97 b}$$

The difference between these possible shake-up configurations is thus the location of the shake-up hole which may be localized in the metal-CO bonding region (Eq. (97 a)) or on the CO molecule itself (Eq. (97 b)).

c) So far, only Loubriel[209)] has performed a first principles calculation of shake-up intensities. Surprisingly enough, he finds a very low probability for metal-CO charge-transfer shake-up (Eq. (97 a)) and a very high intensity for intra-molecular shake-up (Eq. (97 b)). Also, the charge-transfer shake-up satellites are found at too low energies, 2–4 eV above the main line, while the $1\pi \rightarrow 2\pi$ like intra-molecular shake-up satellite, appears at around 6 eV in agreement with experiment. If Loubriel is right, this seems to imply that the CO $1\pi \rightarrow 2\pi$ excitation becomes strongly boosted by simultaneous charge transfer, increasing the 1π–2π overlap.

d) Loubriel's calculation can be criticized in a number of respects:

i) The muffin-tin approximation might give a poor description of the metal (3 d)–CO(2π) bonding region.

ii) There could be coupling between the two shake-up configurations in Eq. (97), perhaps causing transfer of intensity to the charge transfer process.

iii) The calculation did not consider spin-polarization and multiplet splitting effects. In the case of free CO, the scheme does not give a correct picture of the $1\pi \rightarrow 2\pi$ shake-up spectrum, in particular the $2\underline{\sigma}\,^1(1\underline{\pi}2\pi)$ singlet level at around 15 eV. Including spin couplings, the shake-up levels might move in energy and the agreement found by Loubriel with experimental satellite positions might be fortuitous. In this context, one should recall that the charge-transfer satellites in the transition metal complexes discussed in Sect. 8.3 were consistently predicted at too low shake-up energies. A qualitative understanding of the problem must involve simultaneous description of positions and intensities of the 6, 8 and 15 eV satellites in Fig. 67.

e) A detailed treatment of e.g. a $Ni(CO)_4$ molecule using an accurate basis set within the framework of many-body theory seems to be extremely desirable, and should be feasible using the techniques of e.g. Cederbaum and coworkers (cf. Sect. 8.3.3)[209)]. In particular, displacement of the bonding charge has to be treated within the RPAE. Furthermore, the Coulomb interaction between the CO core hole and the shake-up hole in the bond, described by hole-hole scattering processes, most likely has to be treated non-perturbatively since there might be pronounced effects of localization, similar to the case of Ni metal discussed in Sec. 8.3.4.

In conclusion, much interest is presently focussed on the dynamics and excitation spectra of transition-metal carbonyls and adsorbate systems, and I have little doubt that the problems discussed in this section will be understood before long. However, the problem of first principles descriptions of photoionization spectra of these systems will remain a very important field of application of many-body theory for a long time to come[209)].

8.5 Xenon Revisited: Chemical Effects in 4s, 4p X-Ray Photo-Electron Spectra

Since the original emphasis was on core hole spectra in Xe, it is only appropriate that we conclude this review article by returning to the 4s, 4p spectrum in Xe. From the analysis in Sect. 6.1. it is clear that the 4p hole spectrum in atomic Xe represents a border case: If the relative positions of the 4p and $4d^2$ ΔSCF levels could be varied, one could expect to observe significant changes in the spectrum. In Fig. 69 the results of an experimental study of XeF_2 and XeF_4 by Bancroft et al.[210] are compared with the original spectrum of atomic Xe[7,8]. The 4s, 4p levels are shifted by ~3 eV in XeF_2 and ~5.5 eV in XeF_4 to *higher* binding energy relative to atomic Xe, indicating that there is appreciable charge transfer from xenon towards the fluorine ligands. A double vacancy in the n=4 shell will then be shifted by twice as much as a single hole, i.e. about 6 eV in XeF_2 and 11 eV in XeF_4. The resulting single- and double-hole levels have been marked in Fig. 69.

As noted by Bancroft et al.[210], the $4d^2$ levels become pushed to higher binding energy relative to the 4p levels by an amount equal to the chemical shift, leading to a situation reminiscent of that in the elements Cs and Ba with higher nuclear charge Z. In particular, the 4p spectrum of XeF_4 (Fig. 69c) is quite similar to the Cs 4p spectrum in CsI (Fig. 23i), and should resemble even more the spectrum of atomic Cs or, probably still better, metallic Cs.

In a first approximation, the role of the $4d^2$ thresholds is to separate the continuum from the discrete part of the $4d^2$ml levels without really, it seems, influencing the spectral shape. This may be understood in the following way: Since we are dealing with a strongly

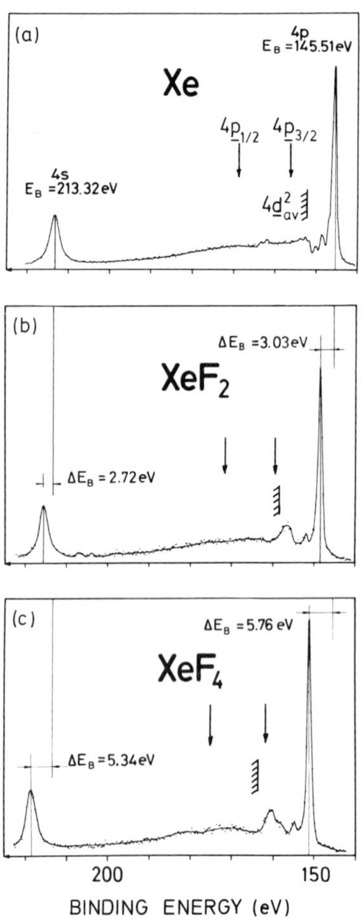

Fig. 69a–c. 4s, 4p X-ray photoelectron spectra of gaseous Xe, XeF_2 and XeF_4 (from[210]). Atomic, relativistic ΔSCF single-[82] and double-hole levels, including chemical shifts, have been marked (for further explanations see text)

interacting system where most of the spectral strength can be associated effectively with a single hole in the n=4 shell, most of the prominent spectral features will have chemical shifts corresponding to a *single* vacancy, not a double vacancy. This means that as we proceed from atomic Xe towards XeF_2 and further on to XeF_4, the spectral density will be approximately constant but the region below the $4d^2$ thresholds will contain a prominent discrete structure that is independent of the double-hole level (Fig. 69). This also means that the relative intensity of the main line should be approximately constant, as should also the relative intensity of the discrete satellite structure when it has fully developed. This picture seems to have become stabilized by Cs (Fig. 23i) and remains unchanged in Ba and La. (Figs. 23k, l). In Ce (Fig. 23 m) the filling of the 4f-levels begin to affect the spectrum, and this effect becomes more and more pronounced in the following elements (Figs. 23 o, p).

An important thing which can be observed in the spectra in Figs. 23 i–m is that as the $4d^2$ thresholds move to higher energy, the $4p_{1/2}$ level starts to recover and give rise to a broad but clearly visible peak structure at around 30 eV higher binding energy than the main peak. This separation is actually predicted by the calculated solutions to the Dyson equation for Xe and Ba in Fig. 25. A trace of $4p_{1/2}$ structure might also be visible in the XeF_4 spectrum (Fig. 69 c).

We shall conclude this review by a comment on open-shell systems, in particular on the development of the 4p-hole spectrum in the rare-earth series, where the 4f shell gradually becomes filled. In this region, the problem reduces to very strong interactions between basically two configurations, $4p\,4f^N$ and $4d^2\,4f^{N+1}$ which, in the spirit of this article, can be expressed as a giant Coster-Kronig (gCK) fluctuation process

$$4p\,4f^N \leftrightarrows 4d^2\,4f^{N+1} \tag{98}$$

This situation is essentially, at hand already in $_{56}$Ba and $_{57}$La (N = 0; Sect. 6.1.1) and also in the analogous case of a 5p hole in Th (Sect. 6.2). As long as the gCK coupling strength (strength of configuration interaction) is large in comparison with multiplet splitting, one can expect the gCK process involving the average configurations in Eq. (98) to give a rough description of the 4p spectral strength distribution. Such a view seems to be supported by experiment because the characteristic shape of the 4p spectrum in $_{56}$Ba and $_{57}$La (Figs. 23 k, l) is clearly visible in the elements $_{58}$Ce to $_{62}$Sm (Figs. 23 m–p). By $_{70}$Yb and $_{71}$Lu the 4f-shell is filled and the positions of the $4p_{1/2,\,3/2}$ levels are accurately given by the ΔSCF method because the gCK fluctuation process is now blocked. I have not seen any experimental results for the region $_{65}$Tb to $_{69}$Tm but I would expect CI, according to Eq. (98), and multiplet splitting to be competing processes. This may lead to significant variations in spectral shape from element to element and to a gradual approach towards the ΔSCF method becoming valid again.

A similar discussion applies to the 4s hole level. However, since the $4s\,4f^N$ and $4p\,4d\,4f^{N+1}$ are energetically quite well separated, one will only observe a strongly shifted "normal" 4s line with intrinsic structure due to exchange splitting (Fig. 23 p). Since the different 4s multiplet levels can become shifted by different amounts, one might be able to observe e.g. exchange narrowing. This is very similar to the case of the 3s core level in MnF_2 discussed by Bagus et al.[211]. It is also closely related to the problem of reduction of exchange splitting in the 3d valence band spectrum in metallic Ni (see e.g.[175, 193]); see also Sect. 8.3.4).

9 Epilogue

Having completed a review article based on the experimental discovery of the missing $4\underline{p}$ ESCA lines in Xe, one is of course bound to find that this kind of effect has been observed experimentally long ago, although not understood, using the much older technique of X-ray emission. Already in the 1920's, Coster and others[212] noted the diffuseness and asymmetry of the $L\gamma_{2,3}$ ($L_1 - N_{2,3}$; $2\underline{s} \rightarrow 4\underline{p}$) X-ray emission spectra (XES) of the elements $_{50}$Sn, $_{52}$Te and $_{53}$I. In 1964, Noreland and Ekstig[212)] produced a detailed $L\gamma_{2,3}$ spectrum which clearly showed a very broad and asymmetric line profile, about 20 eV wide. In 1969 Ekstig and Källne[212)] produced a table of $L\gamma_{2,3}$ widths of the elements $_{40}$Zr to $_{50}$Sn. Very recently, LaVilla[213)] has remeasured the $L\gamma_{2,3}$ XES of $_{50}$Sn, $_{52}$Te and $_{53}$I and the present theory has been applied to XES by Ohno[86)], obtaining excellent agreement with experiment.

Acknowledgement. I should like to thank in particular L. S. Cederbaum, J. W. Davenport, H.-J. Freund, C. K. Jørgensen, G. D. Mahan, M. Ohno, and S. Lundqvist for reading parts of the manuscript, providing useful criticism and stimulating discussions. Many thanks are also due to a large number of people who made valuable comments. I am very grateful to U. Gelius for providing me with experimental results and data analysis. Finally, I should like to thank Department of Physics, Brookhaven National Laboratory, where a large part of this work was done, for their great hospitality. This work has been supported by the Swedish Natural Science Research Council under Contract NFR 2241-029, and by the U. S. Deparment of Energy, Division of Basic Energy Sciences, under Contract No. EY-76-C-02-0016.

10 References

1. Lukirskii, A. P., Zimkina, T. M., Britov, I. A.: Izv. Akad. Nauk. SSSR, Ser. Fiz. *28*, 772 (1964)
2. Codling, K., Madden, R. P.: Appl. Opt. *4*, 1431 (1965)
3. Sonntag, B., Tuomi, T., Zimmerer, G.: Phys. Stat. Sol. *58*, 101 (1972)
4. Haensel, R., Keitel, G., Schreiber, P., Kunz, C.: Phys. Rev. *188*, 1375 (1969)
5. Petersen, H.: DESY internal report F41–73/1 (1973)
   Petersen, H., Radler, K., Sonntag, B., Haensel, R.: J. Phys. B: Atom. Molec. Phys. *8*, 31 (1975); Petersen, H.: Phys. Stat. Sol. (b) *72*, No. 2, 591 (1975)
   Radler, K., Sonntag, B.: DESY report SR-75/20 (1975)
6. Siegbahn, K., Nordling, C., Johansson, G., Hedman, J., Hedin, P. F., Hamrin, K., Gelius, U., Bergmark, T., Werme, L. O., Manne, R., Baer, Y.: ESCA applied to free molecules. Amsterdam: North Holland Publishing Co. 1969
7. Gelius, U.: J. Electron Spectr. *5*, 985 (1974)
8. Svensson, S., Mårtenssen, N., Basilier, E., Malmquist, P. Å., Gelius, U., and Siegbahn, K.: Physica Scripta *14*, 141 (1976)
9. Kowalczyk, S. P., Ley, L., Martin, R. L., McFeely, F. R., Shirley, D. A.: Faraday Discuss. Chem. Soc. *60*, 7 (1975)
10. Kowalczyk, S. P.: Ph. D. thesis 1976. Lawrence Berkeley Laboratory report LBL-4319
11. Fano, U.: Phys. Rev. *124*, 1866 (1961)
12. Anderson, P. W.: Phys. Rev. *124,* 41 (1961)
13. McGuire, E. J.: Phys. Rev. A *9*, 1840 (1974)
14. Wendin, G.: Vacuum ultraviolet radiation physics. Koch, E. E., Haensel, R., Kunz, C. (eds.), p. 225. Braunschweig: Pergamon/Vieweg 1974
15. Lundqvist, S., Wendin, G.: J. Electron Spectr. *5*, 513 (1974)
16. Wendin, G., Ohno, M.: Proceedings of the International Conference on Electron Spectroscopy, June 3–8, Kiev 1975
17. Wendin, G., Ohno, M., Lundqvist, S.: Solid State Commun. *19*, 165 (1976).
18. Wendin, G., Ohno, M.: Proceedings of the 2nd International Conference on Inner Shell Ionisation Phenomena, Freiburg, Germany, March 29/April 2, 1976
19. Wendin, G., Ohno, M.: Physica Scripta *19*, 148 (1976)
20. Jørgensen, C. K.: Structure and Bonding *30*, 141 (1976)
21. Bohm, D., Pines, D.: Phys. Rev. *92*, 609 (1953)
22. Pines, D.: The many-body problem. New York: Benjamin 1961
23. Hedin, L., Lundqvist, S.: Solid State Phys. *23*, 1 (1969)
24. Wendin, G.: The random phase approximation with exchange. In: Photoionization and other probes of many-electron interactions. Wuilleumier, F. (ed.), pp. 61–84. NATO Advanced Study Institute Series. New York: Plenum Press 1976
25. Amusia, M. Ya., Cherepkov, N. A.: Case Studies Atom. Phys. *5*, 47 (1976)
26. Koopmans, T.: Physica *1*, 104 (1934)
27. Nozieres, P.: Theory of interacting fermi systems. New York: Benjamin 1964
    Abrikosov, A., Gorkov, L., Dzyaloshinski, J.: Methods of quantum field theory in statistical physics. Englewood Cliffs: Prentice-Hall 1963
    Fetter, A. L., Walecka, J. D.: Quantum theory of many-particle systems. New York: McGraw-Hill 1971
    Mattuck, R. D.: A guide to Feynman diagrams in the many-body problem, 2nd edn. New York: McGraw-Hill 1976

28. See also Lundqvist[135], Langreth[137], Gadzuk[140], and references cited therein
29. Lundqvist, S.: Int. J. Quantum Chem. *S11*, 379 (1977)
30. Cederbaum, L. S., Domcke, W.: Adv. Chem. Phys. *36*, 295 (1977)
31. Linderberg, J., Öhrn, Y.: Propagators in quantum chemistry. New York: Academic Press 1973
32. Csanak, Gy., Taylor, H. S., Yaris, R.: Adv. Atom. Mol. Phys. *7*, 287 (1971)
33. Manne, R., Åberg, T.: Chem. Phys. Lett. *7*, 282 (1970)
34. Overhauser, A. W.: Phys. Rev. Lett. *4*, 462 (1960)
35. Thouless, D. J.: The quantum mechanics of many-body systems. New York: Academic Press 1961
36. Goldstone, J.: Nuovo Cimento *19*, 15 (1961)
37. Cizek, J., Paldus, J.: J. Chem. Phys. *47*, 3976 (1967) and references cited therein
38. Falicov, L. M., Harris, R. A.: J. Chem. Phys. *51*, 3153 (1969)
39. Snyder, L. C.: J. Chem. Phys. *55*, 95 (1970)
40. Bagus, P. S., Schaeffer, H. F.: J. Chem. Phys. *56*, 224 (1972)
41. Cederbaum, L. S., Domcke, W.: J. Chem. Phys. *66*, 5084 (1977)
Cederbaum, L. S., Domcke, W., Schirmer, J.: Phys. Rev. A *22*, 206 (1980)
42. Martin, R. L., Davidson, E. R.: Chem. Phys. Lett. *51*, 237 (1977)
43. Hedin, L., Johansson, A.: J. Phys. B: Atom. Mol. Phys. *2*, 1336 (1969)
44. Åberg, T.: Shake theory of multiple excitation processes. In: Photoionization and other probes of many-electron interactions. Wuilleumier, F. (ed.), pp. 49–59. NATO Advanced Study Institute Series. New York: Plenum Press 1976
45. Carlson, T. A., Nestor, C. W.: Phys. Rev. A *8*, 2887 (1973)
46. Meldner, H. W., Perez, J. D.: Phys. Rev. A *4*, 1388 (1971)
47. Combescot, M., Nozieres, P.: J. Phys. (France) *32*, 913 (1971)
48. Mahan, G. D.: Solid State Phys. *29*, 75 (1974)
49. Hedin, L.: J. Phys. (France) *39*, C4-103 (1978)
50. Schönhammer, K., Gunnarsson, O.: Solid State Commun. *23*, 691 (1977)
Gunnarsson, O., Schönhammer, K.: Solid State Commun. *26*, 147 (1978)
51. Connerade, J. P., Mansfield, M. W. D., Thimm, K., Tracy, D. H.: Vacuum ultraviolet radiation physics. Koch, E. E., Haensel, R., Kunz, C. (eds.), p. 243. Braunschweig: Pergamon/Vieweg 1974
52. Connerade, J. P., Mansfield, M. W. D., Newsom, G. H., Tracy, D. H., Baig, A., Thimm, K.: Phil. Trans. Roy. Soc. *290*, 327 (1978)
53. Connerade, J. P., Rose, S. J., Grant, I. P.: J. Phys. B *12*, L53 (1979)
54. Mehlhorn, W., Breuckmann, B., Hausamann, D.: Physica Scripta *16*, 177 (1977)
55. Breuckmann, B.: Ph. D. thesis, University of Freiburg 1978
56. Almbladh, C. O.: Phys. Rev. B *16*, 4343 (1977)
57. Mahan, G. D.: Phys. Rev. B *15*, 4387 (1977)
58. Shirley, D. A.: Phys. Rev. A *7*, 1520 (1973)
59. Connerade, J. P.: J. Phys. B: Atom. Mol. Phys. *10*, L239 (1977)
Connerade, J. P.: Proc. Roy. Soc. Lond. A *352*, 561 (1977)
Connerade, J. P.: Proc. Roy. Soc. Lond. A *354*, 511 (1977)
60. Lee, S. T., Süzer, S., Mathias, E., Rosenberg, R. A., Shirley, D. A.: J. Chem. Phys. *66*, 2496 (1977)
61. Veal, B. W., Lam, D. J., Diamond, H., Hoekstra, H. R.: Phys. Rev. B *15*, 2929 (1977)
62. Bancroft, G. M., Sham, T. K., Larsson, S.: Chem. Phys. Lett. *46*, 551 (1977)
63. Krause, M. O., Nestor, C. W.: Physica Scripta *16*, 285 (1977)
64. Kowalczyk, S. P.: Proceedings of the 3rd International Conference on the Electronic Structure of the Actinides, Grenoble August 30/Sept. 1, 1978
65. Sham, T. K., Wendin, G.: Phys. Rev. Lett. *44*, 817 (1980)
66. Ohno, M., Wendin, G.: J. Phys. B: Atom. Mol. Phys. *11*, 1557 (1978)
Ohno, M., Wendin, G.: Solid State Commun. *24*, 75 (1977)
Ohno, M., Wendin, G.: Physica Scripta *16*, 299 (1977)
67. Ohno, M., Wendin, G.: J. Phys. B: Atom. Mol. Phys. *12*, 1305 (1979)
68. Svensson, S., Mårtenssen, N., Basilier, E., Malmqvist, P.-Å., Gelius, U., Siegbahn, K.: J. Electron Spectr. *9*, 51 (1976)

69. Kelly, H. P.: Photoionization cross sections and Auger rates calculated by many-body perturbation theory. In: Photoionization and other probes of many-electron interactions. Wuilleumier, F. (ed.), pp. 83–109. NATO Advanced Study Institute Series. New York: Plenum Press 1976
70. Kadanoff, L. P., Baym, G.: Quantum statistical mechanics. New York: Benjamin 1962
71. Oddershede, J., Jørgensen, P.: J. Chem. Phys. *66,* 1541 (1977)
Oddershede, J., Jørgensen, P., Beebe, N. H. F.: J. Phys. B: Atom. Mol. Phys. *11,* 1 (1978)
72. Pickup, B. T., Goscinski, O.: Mol. Phys. *26,* 1013 (1973)
73. Born, G., Kurtz, A., Öhrn, Y.: J. Chem. Phys. *68,* 74 (1978)
Born, G., Öhrn, Y.: Chem. Phys. Lett. *61,* 307 (1979)
74. Shibuya, T.-I., McKoy, V.: Phys. Rev. A *2,* 2208 (1970)
75. Tsui, F. S. M., Freed, K. F.: Chem. Phys. *14,* 27 (1976)
Herman, M. F., Freed, K. F., Yeager, D. L.: Chem. Phys. *32,* 437 (1978)
76. Garpman, S., Lindgren, I., Lindgren, J., Morrison, J.: Phys. Rev. A *11,* 758 (1975)
Lindgren, I., Morrison, J.: Atomic many-body theory. Springer Verlag, Chem. Phys., in press
77. Chang, T. N., Fano, U.: Phys. Rev. A *13,* 263, 282 (1976)
78. Starace, A. F., Armstrong Jr., L.: Phys. Rev. A *13,* 1850 (1976)
79. Brown, E. R., Carter, S. L., Kelly, H. P.: Phys. Lett. *66A,* 290 (1978)
80. Desclaux, J. P.: At. Data Nucl. Data Tables *12,* 311 (1973)
81. Wendin, G.: Phys. Lett. 51 A, 291 (1975)
82. Huang, K. N., Aoyagi, M., Chen, M. H., Craseman, B.: At. Data Nucl. Data Tables *18,* 243 (1976)
83. Johansson, B., Mårtensson, N.: Phys. Rev., B *21,* 4427 (1980)
Mårtensson, N., Johansson, B.: Solid State Commun. *32,* 791 (1979)
84. Chen, M. H., Crasemann, B., Yin, L., Tsang, T., Adler, I.: Phys. Rev. A *13,* 1435 (1976)
85. Ohno, M., Wendin, G.: to be published
86. Ohno, M.: J. Phys. C: Solid State Phys. *13,* 447 (1980); Physica Scripta *21,* 589 (1980)
87. Wendin, G.: J. Phys. B: Atom. Mol. Phys. *9,* L297 (1976)
88. Wendin, G., Starace, A. F.: J. Phys. B: Atom. Mol. Phys. *11,* 4119 (1978)
89. Aoyagi, M., Chen, M. H., Crasemann, B., Huang, K.-N., Mark, H.: Abstracts of contributed papers. p. 301, 9th ICPEAC, Seattle, 1975
90. Newns, D. M.: Phys. Rev. *178,* 1123 (1969)
91. Lundqvist, B. I., Hjelmberg, H., Gunnarsson, O.: Adsorbate induced electronic states. In: Photoemission and the electronic properties of surfaces. Feuerbacher, B., Fitton, B., Willis, R. F. (eds.). New York: John Wiley-Interscience 1978
92. Cohen-Tannoudji, C., Avan, P.: Discrete states coupled to a continuum: Continuous transition between Rabi precession and the Weisskopf-Wigner exponential decay. In: C.N.R.S. Colloquium Etats Atomiques et Moléculaires Couples à un Continuum, p. 93. Aussois, France. Paris: Editions du CNRS no. 273
93. Connerade, J. P.: Proc. Roy. Soc. A *362,* 361 (1978)
94. Wendin, G.: Physica Scripta *16,* 296 (1977)
95. Adam, M.Y., Wuilleumier, F., Sander, N., Schmidt, V., Wendin, G.: J. Phys. (France) *39,* 129 (1978)
96. Minnhagen, L.: Ark. Fys. *21,* 415 (1962); *25,* 203 (1963)
97. Wertheim, G. K., Rosencweig, A.: Phys. Rev. Lett. *26,* 1179 (1971)
98. Luyken, B. F. J.: Physica *60,* 432 (1972)
Yafet, Y., Watson, R. E.: Int. J. Quant. Chem. *S 7,* 93 (1973)
99. McCarthy, I. E., Weigold, E.: Phys. Repts. *27 C,* 277 (1976)
100. Martin, R. L., Kowalczyk, S. P., Shirley, D. A.: Lawrence Berkeley Laboratory Report LBL-5445 (1977)
101. Adam, M. Y., Wuilleumier, F., Krummacher, S., Schmidt, V., Mehlhorn, W.: J. Phys. B: Atom. Mol. Phys. *12,* L 1 (1978);
Adam, M. Y., Wuilleumier, F., Krummacher, S., Sandner, S., Schmidt, V., Mehlhorn, W.: J. Electron Spectr. *15,* 211 (1979)
102. Hansen, J. E., Persson, W.: Phys. Rev. A *18,* 1459 (1978)
103. Dyall, K. G., Larkins, F. P.: J. Electron Spectr. *15,* 165 (1978)
104. Cowan, R. D., Radziemski Jr., L. J., Kaufman, V.: J. Opt. Soc. Am. *64,* 1474 (1974)

105. Reader, G. L., Epstein, J.: J. Opt. Soc. Am. *66,* 590 (1976)
106. Hansen, J. E.: J. Opt. Soc. Am. *67,* 754 (1977)
107. Wendin, G.: Many-electron effects in photoionisation. In: Photoionisation of atoms and molecules. Buckley, B. D. (ed.), pp. 1–21. Daresbury Laboratory report DL/SCI/R 11 (Atomic and Molecular) 1978
108. Rose, S. J., Grant, I. P., Connerade, J. P.: Phil. Trans. Roy. Soc. *296,* 527 (1980)
109. Schirmer, J., Cederbaum, L. S., Domcke, W., von Niessen, W.: Chem. Phys. *26,* 149 (1978)
110. Schirmer, J., Domcke, W., Cederbaum, L. S., von Niessen, W.: J. Phys. B: Atom. Mol. Phys. *11,* 1901 (1978)
111. Cederbaum, L. S., Domcke, W., Schirmer, J., von Niessen, W., Diereksen, G. H. F., Kraemer, W. P.: J. Chem. Phys. *69,* 1591 (1978)
112. Bagus, P. S., Viinikka, E.-K.: Phys. Rev. A *15,* 1486 (1977)
113. Shirley, D. A.: Adv. Chem. Phys. *23,* 85 (1973)
114. Basch, H.: J. Electron Spectr. *5,* 463 (1974)
115. This model case, with the application to ethylene, has recently been discussed by S. Lundqvist, Bull. Am. Phys. Soc. *22,* 437 (1977) (invited paper) and at the Colloquium on the Chemical Physics of Surfaces, Catalysis and Membranes, Uppsala, 1977 (invited paper)
116. Lundqvist, B. I.: Phys. Kond. Materie *7,* 117 (1968)
117. Cederbaum, L. S., Schirmer, J., Domcke, W., von Niessen, W.: Int. J. Quant. Cem. *15,* 593 (1978); Schirmer, J., Domcke, W., Cederbaum, L. S., von Niessen, W., Åsbrink, L.: Chem. Phys. Lett. *61,* 30 (1979)
118. Cederbaum, L. S., Domcke, W., Schirmer, J., von Niessen, W.: Physica Scripta *21,* 481 (1980)
119. Cederbaum, L. S., Schirmer, J., Domcke, W., von Niessen, W.: J. Phys. B: Atom. Mol. Phys. *10,* L 549 (1977)
120. Rose, J., Shibuya, T.-I., McKoy, V.: J. Chem. Phys. *58,* 74 (1973)
121. It should be noted that the relevant approximation scheme for the self-energy depends on whether one considers core or inner-valence holes or outer-valence holes. In the latter case, Fermi sea correlations (ground-state correlations) are relatively more important than relaxation and screening which, as a reasonable approximation, can be neglected in the intermediate states in the self-energy. Instead, it is important to include the ground-state correlation part (Fig. 9 g) when solving the Dyson equation (Eq. (15)), which then leads to the inclusion of certain types of higher excitations. Cederbaum and coworkers have found this procedure to give accurate results for outer-valence levels. On the other hand, for inner-valence (or core) levels, relaxation and screening in the intermediate 2 h-1 p levels are very important and have to be considered, as discussed above
122. Dill, D., Siegel, J., Wallace, S., Dehmer, J. L.: Phys. Rev. Lett. *41,* 1230 (1978); erratum, ibid. *42,* 411 (1979)
123. Cederbaum, L. S.: Chem. Phys. Lett. *25,* 562 (1974)
124. Gelius, U., Basilier, E., Svensson, S., Bergmark, T., Siegbahn, K.: J. Electron Spectr. *2,* 405 (1973)
125. Snyder, L. C., Basch, H.: Molecular wave functions and properties. New York: Wiley-Interscience 1972
126. In[7] this satellite is erroneously ascribed an intensity of 0.25 of the main line instead of $\lesssim 0.10$, and this error seems to have propagated through recent literature. T. A. Carlson (private commun. 1978) has estimated the satellite strength to be ~ 0.075[130]
127. Hillier, I. H., Kendrick, J.: J. Electron Spectr. *2,* 405 (1976)
128. Weigold, E., Dey, S., Dixon, A. J., McCarthy, I. E., Lassey, K. R., Teubner, P. J. O.: J. Electron Spectr. *10,* 177 (1977)
129. Cavell, R. G., Allison, D. A.: J. Chem. Phys. *69,* 159 (1978)
130. Carlson, T. A., Dress, W. B., Grimm, F. A., Haggerty, J. S.: J. Electron Spectr. *10,* 147 (1977)
131. Carlson, T. A., Krause, M. O., Moddeman, W. E.: J. Phys. (France) *32,* C4–76 (1971)
132. Friedel, J.: Philos. Mag. *43,* 153 (1952); Adv. Phys. *3,* 446 (1954)
133. Mahan, G. D.: Phys. Rev. *163,* 612 (1967)
134. Nozières, P., de Dominis, C. I.: Phys. Rev. *178,* 1097 (1969)
135. Lundqvist, B. I.: Phys. Kond. Materie *9,* 236 (1969)
136. Friedel, J.: Comments Solid State Phys. *2,* 21 (1969)

137. Langreth, D. C.: Phys. Rev. B *1,* 471 (1970)
Langreth, D. C.: Theory of plasmon effects in high energy spectroscopy. In: Collective properties of physical systems. Lundqvist, B., Lundqvist, S. (eds.), pp. 210–222. Nobel Symposium 24. Stockholm. The Nobel Foundation, Academic Press 1973
138. Doniach, S., Sunjić, M.: J. Phys. C: Solid State Phys. *3,* 285 (1970)
139. Wertheim, G. K.: Jap. J. Appl. Phys. *17,* Suppl. 17-2, 33 (1978)
140. Gadzuk, J. W.: Many-body effects in photoemission. In: Photoemission and the electronic properties of surfaces. Feuerbacher, B., Fitton, B., Willis, R. F. (eds.). New York: Wiley-Interscience 1978
141. Ichii, T., Sakisaka, Y., Yamaguchi, S., Hanyu, T., Ischii, H.: J. Phys. Soc. Japan *42,* 876 (1977)
142. von Barth, U., Grossman, G.: Solid State Commun. *32,* 695 (1980)
143. Swarts, C. A., Dow, J. D., Flynn, C. P.: Phys. Rev. Lett. *43,* 158 (1979)
144. Nougera, C., Spanjaard, D., Friedel, D.: J. Phys. F: Metal Phys. *9,* 1189 (1979)
145. Kotani, A., Toyozawa, Y.: J. Phys. Soc. Japan *37,* 912 (1974); Kotani, A.: Jap. J. Appl. Phys. *17,* Suppl. 17-2, 243 (1978)
146. Kotani, A., Toyozawa, Y.: J. Phys. Soc. Japan *37,* 563 (1974)
147. Gunnarsson, O., Schönhammer, K.: Surface Sci. *80,* 471 (1979)
148. Wendin, G.: Int. J. Quant. Chem. *S 13,* 659 (1979)
149. Ley, L., Kowalczyk, S. P., McFeely, F. R., Pollak, R. A., Shirley, D. A.: Phys. Rev. B *8,* 2392 (1973)
150. Kim, K. S.: Chem. Phys. Lett. *29,* 234 (1974); J. Electron Spectr. *3,* 217 (1974); Phys. Rev. B *11,* 2177 (1975)
151. Wallbank, B., Main, I. G., Johnson, C. E.: J. Electron Spectr. *5,* 259 (1974) and references cited therein
152. Vernon, G. A., Stucky, G., Carlson, T. A.: Inorg. Chem. *15,* 278 (1976) and references cited therein
153. Watson, R. E., Perlman, M.: Structure and Bonding *24,* 83 (1975) and references cited therein
154. Wallbank, B., Perera, J. S. H. Q., Frost, D. C., Mc Dowell, C. A.: J. Chem. Phys. *69,* 5405 (1975)
155. Larsson, S.: Chem. Phys. Lett. *32,* 401 (1975)
156. Larsson, S.: J. Electron Spectr. *8,* 171 (1971)
157. Larsson, S.: Physica Scripta *16,* 378 (1977); *21,* 558 (1980)
158. Wertheim, G. K., Cohen, R. L., Rosencwaig, A., Guggenheim, H. J.: In: Electrospectroscopy. Shirley, D. A. (ed.), p. 813. Amsterdam: North-Holland 1972
159. Jørgensen, C. K., Berthou, H.: Chem. Phys. Lett. *13,* 186 (1972)
160. Signorelli, A. J., Hayes, R. G.: Phys. Rev. B *8,* 81 (1973)
161. Suzuki, S., Ishii, T., Sagawa, T.: J. Phys. Soc. Japan *37,* 1334 (1974)
162. Berthou, H., Jørgensen, C. K., Bonnelle, C.: Chem. Phys. Lett. *38,* 199 (1976)
163. Weber, J., Berthou, H., Jørgensen, C. K.: Chem. Phys. Lett. *45,* 1 (1977)
164. Wallbank, B., Sham, T. K., Esquivel, J. L., Larsson, S.: Chem. Phys. Lett. *51,* 105 (1977)
165. Lang, N. D., Williams, A. R.: Phys. Rev. B *16,* 2408 (1977)
166. Hoogewijs, R., Vennik, J.: Solid State Commun. *31,* 531 (1979)
167. Hussain, S. S., Newns, D.: Solid State Commun. *25,* 1049 (1978)
168. Gadzuk, J. W., Doniach, S.: Surface Sci. *77,* 427 (1978)
169. Gunnarsson, O., Schönhammer, K.: Phys. Rev. Lett. *41,* 1608 (1978)
170. Gumhalter, B.: Surface Sci. *80,* 459 (1979)
171. Plummer, E. W., Salaneck, W. R., Miller, J. J.: Phys. Rev. B *18,* 1673 (1978)
172. Wertheim, G. K., Campagna, M.: Solid State Commun. *26,* 553 (1978)
173. Crecelius, C., Wertheim, G. K., Buchanan, D. N. E.: Phys. Rev. B *18,* 6519 (1978)
174. Himpsel, F. J., Eastman, D. E., Koch, E. E., Williams, A. R.: Phys. Rev. B *22,* 4604 (1980)
175. Eastman, D. E., Janak, J. F., Williams, A. R., Coleman, R. V., Wendin, G.: J. Appl. Phys. *50,* 7423 (1979)
176. Pireaux, J. J., Svensson, S., Basilier, E., Malmqvist, P.-Å., Gelius, U., Caudano, R., Siegbahn, K.: Phys. A *14,* 2133 (1976)
177. Pireaux, J. J., Caudano, R.: Phys. Rev. B *15,* 2242 (1977)
178. Pireaux, J. J., Riga, J., Caudano, R., Verbist, J. J.: Physica Scripta *16,* 329 (1977)

179. Duke, C. B., Paton, A., Salaneck, W. R., Thomas, H. R., Plummer, E. W., Heeger, A. J., MacDiarmid, A. G.: Chem. Phys. Lett. *59,* 146 (1978)
180. Salaneck, W. R., Lin, J. W-p., Epstein, A. J.: Phys. Rev. B *13,* 5574 (1976) and references cited therein
181. Rosencwaig, A., Wertheim, G. K., Guggenheim, H. J.: Phys. Rev. Lett. *27,* 479 (1971)
182. Baer, Y., Hauger, R., Zuercherer, Ch., Campagna, M., Wertheim, G. K.: J. Electron Spectr. *15,* 27 (1979)
183. Kanski, J., Nilsson, P. O., Curelaru, I.: J. Phys. F: Metal Phys. *6,* 1073 (1976)
184. Wendin, G., Nuroh, K.: Phys. Rev. Lett. *39,* 48 (1977)
185. Tossel, J. A.: J. Electron Spectr. *10,* 169 (1977)
186. Braga, M., Larsson, S.: Int. J. Quant. Chem. *S11,* 61 (1977)
187. Kemeny, P. C., Shevchik, N. J.: Solid State Commun. *17,* 255 (1974)
188. Hüfner, S., Wertheim, G. K.: Phys. Lett. *51A,* 299 (1975)
189. Hüfner, S., Wertheim, G. K.: Phys. Lett. *51A,* 301 (1975)
190. Eberhardt, W., Plummer, E. W.: Phys. Rev. B *21,* 3245 (1980)
191. Iwan, M., Himpsel, F., Eastman, D. E.: Phys. Rev. Lett. *43,* 1829 (1979)
192. Grebennikov, V. I., Babanov, Yu. A., Sokolov, O. B.: Phys. Stat. Solidii (b) *79,* 423 (1977); *80,* 73 (1977)
193. Davis, L. C., Feldkamp, L. A.: J. Appl. Phys. *50,* 1944 (1979)
194. Yin, L., Tsang, T., Adler, I.: Phys. Rev. B *15,* 2975 (1977)
195. Antonides, E., Jantze, E. C., Sawatsky, G. A.: Phys. Rev. B *15,* 1669 (1977)
196. Sawatsky, G. A.: Phys. Rev. Lett. *39,* 504 (1977)
197. Cini, M.: Solid State Commun. *24,* 681 (1977)
198. Penn, D. R.: Phys. Rev. Lett. *42,* 921 (1979)
199. Liebsch, A.: Phys. Rev. Lett. *42,* 921 (1979)
200. Recently, Davis, L. C., Feldkamp, L. A.: Solid State Commun. *34,* 141 (1980) have obtained excellent agreement with the experimental valence band XPS spectrum, including the 6 eV satellite, by using the self-energy of a localized 3p-hole to approximately represent the self-energy of a 3d-hole and to evaluate the effects on the 3d-band structure. This is in agreement with the discussion in ref.[175]. See also Feldkamp, L. A., Davis, L. C.: Phys. Rev. B. *22,* 3644 (1980); *22,* 4994 (1980), and Treglia, G., Ducastelle, F., Spanjaard, D.: Phys. Rev. B *21,* 3729 (1980) for detailed discussions and further references
201. Fuggle, J. C., Umbach, E., Menzel, D., Wandelt, K., Brundle, C. R.: Solid State Commun. *27,* 65 (1978)
202. Bancroft, G. M., Boyd, B. D., Creber, D. K.: Inorg. Chem. *17,* 1008 (1978)
203. Ellis, D. E., Baerends, E. J., Adachi, H., Averill, F. W.: Surface Sci. *64,* 649 (1977)
204. Baerends, E. J., Ros, P.: Int. J. Quant. Chem. *S12,* 169 (1978)
205. Loubriel, G.: Phys. Rev. B *20,* 5339 (1979)
206. Domcke, W., Cederbaum, L. S., Schirmer, J., von Niessen, W.: Phys. Rev. Lett. *42,* 1237 (1979)
207. An experimental $Ni(CO)_4$ spectrum has not yet been published (Brundle, C. R., to be published) but appears to be quite similar to that of $Cr(CO)_6$, which we therefore use as an illustration (Fig. 67).
208. Kim, B.-I., Adachi, H., Imoto, S.: J. Electron Spectr. *11,* 349 (1977)
209. After this article was completed, H.-J. Freund has drawn my attention to some recent theoretical work that I was not aware of. In the case of free CO, Rodwell, W. R., Guest, M. F., Darko, T., Hillier, I. H., Kendrick, J.: Chem. Phys. *22,* 467 (1977) have correctly treated the $1\pi \rightarrow 2\pi$ shake-up excitations in C(1 s) ionization, and also discussed assignments of the higher-lying shake-up levels (cf. Section 7.3). In the case of transition-metal (TM) carbonyls and CO-TM adsorbate systems, Bagus, P. S., Herman, K.: Surf. Sci. *89,* 588 (1979) have calculated ground and excited levels of NiCO, and Mitcheson, G. R., Hillier, I. H.: J. Chem. Soc. (Farad. Trans. II) *75,* 929 (1979) the same for $Ni(CO)_4$. The problem of shake-down satellites in the core level spectrum of para-nitroaniline[206] has recently been studied theoretically by Freund, H.-J. and Bigelow, R. W.: Chem. Phys. (in press) both for the vapour and condensed phases. An extensive discussion of TM-carbonyl and TM-CO adsorbate spectra has been given by Freund, H.-J., Plummer, E. W.: Phys. Rev. B (in press) (see this paper for further references). Among other things the authors conclude that a correct treatment of the

$1\pi \rightarrow 2\pi$ intra-CO shake-up levels leads to an interpretation of the main line and intense, low-lying satellite in terms of metal-CO charge transfer and shake-up in the metal-CO bonding region. This corroborates the conclusions of the qualitative discussion in Section 8.4 and speaks against the interpretation of Loubriel[205]. Finally, many-body theory and Green's function techniques have indeed been applied recently to a description of core and valence spectra of TM-carbonyls and adsorbate systems by Saddei, D., Freund, H.-J., Hohlneicher, G.: Surf. Sci. *95,* 527 (1980); Chem. Phys. (in press)

210. Bancroft, G. M., Malmqvist, P.-Å., Svensson, S., Basilier, E., Gelius, U., Siegbahn, K.: Inorg. Chem. *17,* 1595 (1978)
211. Bagus, P. S., Freeman, A. J., Sasaki, F.: Phys. Rev. Lett. *30,* 850 (1973).
212. For a discussion of the history of the problem and for further and detailed references, see ref.[213]
213. LaVilla, R. E.: Phys. Rev. A *17,* 1018 (1978)
214. Boring, M., Cowan, R. D., Martin, R. L.: Phys. Rev. B *23,* 445 (1981)

Author-Index Volumes 1–45

Lecture Notes in Chemistry

Volume 1: G. H. Wagnière: **Introduction to Elementary Molecular Orbital Theory and to Semiempirical Methods.** 1976. 33 figures. V, 109 pages ISBN 3-540-07865-7

Volume 2: E. Clementi: **Determination of Liquid Water Structure.** Coordination Numbers for Ions and Solvation for Biological Molecules. 1976. 32 figures, 18 tables. VI, 107 pages ISBN 3-540-07870-3

Volume 3: S. R. Niketić, K. Rasmussen: **The Consistent Force Field.** A Documentation. 1977. 24 figures, 26 tables. IX, 212 pages. ISBN 3-540-08344-8

Volume 4: A. Graovac, I. Gotman, N. Trinajstić. **Topological Approach to the Chemistry of Conjugated Molecules.** 1977. 34 figures, 2 tables. IX, 123 pages. ISBN 3-540-08431-2

Volume 5: R. Carbó, J. M. Riera. **A General SCF Theory.** 1978. 25 figures, 19 tables. XII, 210 pages ISBN 3-540-08535-1

Volume 6: I. Hargittai: **Sulphone Molecular Structures.** Conformation and Geometry from Electron Diffraction and Microwave Spectroscopy; Structural Variations. 1978. 40 figures. VII, 175 pages ISBN 3-540-08654-4

Volume 7: **Ion Cyclotron Resonance Spectrometry.** Editors: H. Hartmann, K.-P. Wanczek. 1978. 66 figures, 32 tables. V, 326 pages ISBN 3-540-08760-5

Volume 8: E. E. Nikitin, L. Zülicke: **Selected Topics of the Theory of Chemical Elementary Processes.** 1978. 41 figures, 6 tables. IX, 175 pages ISBN 3-540-08768-0

Volume 9: A. Julg: **Crystals as Giant Molecules.** 1978. 8 figures, 21 tables. VII, 135 pages ISBN 3-540-08946-2

Volume 10: J. Ulstrup: **Charge Transfer Processes in Condensed Media.** 1979. 32 figures. VII, 419 pages. ISBN 3-540-09520-9

Volume 11: F. A. Gianturco: **The Transfer of Molecular Energies by Collision: Recent Quantum Treatments.** 1979. 17 figures, 6 tables. VIII, 327 pages. ISBN 3-540-09701-5

Volume 12: **The Permutation Group in Physics and Chemistry.** Editor: J. Hinze. 1979. 20 figures, 13 tables. VI, 230 pages. ISBN 3-540-09707-4

Volume 13: G. Del Re, G. Berthier, J. Serre: **Electronic States of Molecules and Atom Clusters.** Foundations and Prospects of Semiempirical Methods. 1980. VIII, 177 pages. ISBN 3-540-09738-4

Volume 14: E. W. Thulstrup: **Aspects of the Linear and Magnetic Circular Dichroism of Planar Organic Molecules.** 1980. 27 figures, 5 tables. VI, 100 pages. ISBN 3-540-09754-6

Volume 15: **Steric Fit in Quantitative Structure-Activity Relations.** By A. T. Balaban, A. Chiriac, I. Motoc, Z. Simon. 1980. 20 figures, 43 tables. VII, 178 pages. ISBN 3-540-09755-4

Volume 16: P. Cársky, M. Urban: **Ab Initio Calculations.** Methods and Applications in Chemistry. 1980. 23 figures, 54 tables. VI, 247 pages ISBN 3-540-10005-9

Volume 17: H. G. Hertz: **Electrochemistry.** A Reformulation of the Basic Principles. 1980. 44 figures. X, 254 pages. ISBN 3-540-10008-3

Volume 18: S. G. Christov: **Collision Theory and Statistical Theory of Chemical Reactions.** 1980. 30 figures, 9 tables. XII, 322 pages ISBN 3-540-10012-1

Volume 19: E. Clementi: **Computational Aspects for Large Chemical Systems.** 1980. 78 figures, 59 tables. V, 184 pages. ISBN 3-540-10014-8

Springer-Verlag
Berlin
Heidelberg
New York

A. F. Williams

A Theoretical Approach to Inorganic Chemistry

1979. 144 figures, 17 tables. XII, 316 pages
ISBN 3-540-09073-8

This book is intended to outline the application of simple quantum mechanics to the study of inorganic chemistry, and to show its potential for systematizing and understanding the structure, physical properties, and reactivities of inorganic compounds. The considerable development of inorganic chemistry in recent years necessitates the establishment of a theoretical framework if the student is to acquire sound knowledge of the subject. An effort has been made to cover a wide range of subjects, and to encourage the reader to think of further extensions of the theories discussed. The importance of the critical application of theory is emphasized, and, although the book is concerned chiefly with molecular orbital theory, other approaches are discussed. The book is intended for students in the latter half of their undergraduate studies.

Contents: Quantum Mechanics and Atomic Theory. – Simple Molecular Orbital Theory. – Structural Applications of Molecular Orbital Theory. – Electronic Spectra and Magnetic Properties of Inorganic Compounds. – Alternative Methods and Concepts. – Mechanism and Reactivity. – Descriptive Chemistry. – Physical and Spectroscopic Methods. – Appendices. – Subject Index.

Springer-Verlag
Berlin
Heidelberg
New York